Udo Schüklenk

Access to Experimental Drugs in Terminal Illness
Ethical Issues

Pre-publication
REVIEWS,
COMMENTARIES,
EVALUATIONS . . .

"The best book yet on this subject—a thorough and devastating critique of the paternalistic position that people infected with HIV should not be able to take risks with dangerous, experimental treatments. Written with panache and passion. Highly recommended."

Gregory Pence, PhD
School of Medicine
and Department of Philosophy,
University of Alabama
at Birmingham

"It is unusual these days for scholars to straddle disciplines, yet Schüklenk does this consummately. He marshalls empirical evidence and ethical theory to produce a compelling case. His conclusion deserves the most careful consideration by scientists and policy-makers alike—but I suspect the book will be of greatest help to the presently excluded community it seeks to serve."

David Seedhouse, PhD
Editor, *Health Care Analysis:*
Journal of Health Philosophy and Policy,
School of Medicine,
University of Auckland,
New Zealand

More pre-publication
REVIEWS, COMMENTARIES, EVALUATIONS . . .

"This book is both a work of rigorous philosophy and a devastating exposé of the ethics of many current practices in the regulation of experimental drugs for terminally ill patients. Dr. Schüklenk demolishes all the arguments for restricting experimental drugs that terminally ill patients may wish to try in order to save their lives. More startling still, Dr. Schüklenk shows that drug regulators have knowingly restricted the availability of experimental drugs in order to ensure an adequate supply of 'volunteers' for clinical trials. Thus people with terminal illnesses are effectively being coerced into becoming research subjects.

Dr. Schüklenk has written a work of great practical importance. He forces us to rethink our attitudes toward drug regulation."

Peter Singer, PhD
Professor,
Centre for Human Bioethics,
Monash University,
Clayton, Australia

"This book comments on the ethical and practical problems that emerge in the medical treatment of people with AIDS. While testing the effectiveness of the medication, it proved difficult to construct two samples because one was denied the new drug. When patients are terminally ill, they tend to try any remedy that might promise help—regardless of the risk involved.

The issues discussed in the book are of a general relevance, beyond the actual cause of AIDS. Too seldom are ethical problems analyzed in a scientific way; they are often decided in an unprofessional manner. Modern philosophy can contribute much to overcome simplistic solutions.

Udo Schüklenk does an admirable job of presenting the ethical questions in a precise and sound manner. He disposes of many examples taken from the history of medicine in order to make the logical structure transparent, even for those who are not trained in logics or meta-ethics. He has written a highly readable text that should be of great interest to the specialists in the field and to the general public.

The results of the book are relevant for medical practice as well as for the politics of pharmacology. Schüklenk presents answers that are based on ethical theory; they will help to develop a humanistic solution to the challenges created by terminal illness."

Dr. Rüdiger Lautmann
Professor of Sociology,
University of Bremen,
Germany

NOTES FOR PROFESSIONAL LIBRARIANS AND LIBRARY USERS

This is an original book title published by Pharmaceutical Products Press, an imprint of The Haworth Press, Inc. Unless otherwise noted in specific chapters with attribution, materials in this book have not been previously published elsewhere in any format or language.

CONSERVATION AND PRESERVATION NOTES

All books published by The Haworth Press, Inc. and its imprints are printed on certified pH neutral, acid free book grade paper. This paper meets the minimum requirements of American National Standard for Information Sciences–Permanence of Paper for Printed Material, ANSI Z39.48-1984.

Access to Experimental Drugs in Terminal Illness
Ethical Issues

PHARMACEUTICAL PRODUCTS PRESS
Pharmaceutical Sciences
Mickey C. Smith, PhD
Executive Editor

New, Recent, and Forthcoming Titles:

Pharmaceutical Marketing: Strategy and Cases by Mickey C. Smith

International Pharmaceutical Services: The Drug Industry and Pharmacy Practice in Twenty-Three Major Countries of the World edited by Richard N. Spivey, Albert I. Wertheimer, and T. Donald Rucker

A Social History of the Minor Tranquilizers: The Quest for Small Comfort in the Age of Anxiety by Mickey C. Smith

Marketing Pharmaceutical Services: Patron Loyalty, Satisfaction, and Preferences edited by Harry A. Smith and Joel Coons

Nicotine Replacement: A Critical Evaluation edited by Ovide F. Pomerleau and Cynthia S. Pomerleau

Interpersonal Communication in Pharmaceutical Care by Helen Meldrum

Searching for Magic Bullets: Orphan Drugs, Consumer Activism, and Pharmaceutical Development by Lisa Ruby Basara and Michael Montagne

The Honest Herbal by Varro E. Tyler

Understanding the Pill: A Consumer's Guide to Oral Contraceptives by Greg Juhn

Pharmaceutical Chartbook, Second Edition edited by Abraham G. Hartzema and C. Daniel Mullins

The Handbook of Psychiatric Drug Therapy for Children and Adolescents by Karen A. Theesen

Children, Medicines, and Culture edited by Patricia J. Bush, Deanna J. Trakas, Emilio J. Sanz, Rolf L. Wirsing, Tuula Vaskilampi, and Alan Prout

Social and Behavioral Aspects of Pharmaceutical Care edited by Mickey C. Smith and Albert I. Wertheimer

Studies in Pharmaceutical Economics edited by Mickey C. Smith

Drugs of Natural Origin: Economic and Policy Aspects of Discovery, Development, and Marketing by Anthony Artuso

Pharmacy and the U.S. Health Care System, Second Edition edited by Jack E. Fincham and Albert I. Wertheimer

Medical Writing in Drug Development: A Practical Guide for Pharmaceutical Research by Robert J. Bonk

Improving the Quality of the Medication Use Process: Error Prevention and Reducing Adverse Drug Events edited by Alan Escovitz, Dev S. Pathak, and Philip J. Schneider

Access to Experimental Drugs in Terminal Illness: Ethical Issues by Udo Schüklenk

Herbal Medicinals: A Clinician's Guide by Lucinda G. Miller and Wallace J. Murray

Tyler's Herbs of Choice: The Therapeutic Use of Phytomedicinals by James E. Robbers and Varro E. Tyler

Access to Experimental Drugs in Terminal Illness
Ethical Issues

Udo Schüklenk

Pharmaceutical Products Press
An Imprint of the Haworth Press, Inc.
New York • London

Published by

Pharmaceutical Products Press®, an imprint of The Haworth Press, Inc., 10 Alice Street, Binghamton, NY 13904-1580

© 1998 by The Haworth Press, Inc. All rights reserved. No part of this work may be reproduced or utilized in any form or by any means, electronic or mechanical, including photocopying, microfilm and recording, or by any information storage and retrieval system, without permission in writing from the publisher. Printed in the United States of America.

Cover design by Jennifer M. Gaska.

Library of Congress Cataloging-in-Publication Data

Schüklenk, Udo.
   Access to experimental drugs in terminal illness : ethical issues / Udo Schüklenk.
     p. cm.
   Includes bibliographical references and index.
   ISBN 0-7890-0563-8 (alk. paper)
   1. Terminal care—Moral and ethical aspects. 2. Drugs—Testing—Moral and ethical aspects. 3. Health services accessibility. I. Title.
R726.8.S37 1998
174'.28—DC21
                                                                                                          98-4689
                                                                                                                 CIP

To the memory of two AIDS activists par excellence,
Andreas Salmen and Mike Callen

ABOUT THE AUTHOR

Udo Schüklenk, PhD, is Lecturer in Applied Ethics and Program Director of the MA Bioethics Program at the University of Central Lancashire Centre for Professional Ethics, Preston, United Kingdom. He has taught bioethics and philosophy at universities in Germany and Australia. Dr. Schüklenk has contributed a large number of articles to reference works, such as the *Encyclopedia of Applied Ethics* and *Companion to Bioethics,* and refereed journals, including *Hastings Center Report, Bioethics, Cambridge Quarterly of Healthcare Ethics, Journal International de Bioethique,* and *Ethick in der Medizin.* He serves as an Expert Advisor to the UK National Health Service Executive North West's Multicentre Research Ethics Committee and as a member of the editorial boards of *Bioethics* and *Health Care Analysis.*

CONTENTS

Preface	ix
Introduction	1
Chapter 1. Autonomy—Access to Experimental Drugs and the Terminally Ill	9
Immanuel Kant	15
Weak Paternalism: John Stuart Mill	24
Weak Paternalism: Joel Feinberg	41
Weak Paternalism: Gerald Dworkin	56
Weak Paternalism: Tom L. Beauchamp, James F. Childress, and Ruth R. Faden	64
Strong Paternalism: Robert Young	74
The Importance of Respecting Individual Autonomy	81
Summary of the Philosophical Debate	85
Practical Implications of the Arguments	87
Chapter 2. Should We Restrict Access to Experimental Drugs to Promote Clinical Trials?	89
Research Clinical Trials—Are They Designed to Help Current or Future Patients?	97
Physicians and Patients	101
The Fischl/Richman AZT Trial	107
Ethical Analysis of the Fischl/Richman AZT Trial	121
Coercive Offers, or You Cannot Have It Both Ways	133
Placebos and Other Drugs	138
Summary	140
Chapter 3. Costs and Other Practical Problems	145
Access to Experimental Drugs in Societies with Universal Health Care	151
Access to Experimental Drugs in Societies Without Universal Health Care	156
Provision of Information About Experimental Drugs	161
Practical Cultural Hindrances in Exercising Autonomy	167

Conclusion	169
Notes	173
Bibliography	205
Index	219

Preface

The questions addressed in this book concern problems that have concerned AIDS activists in most Western countries over the last fifteen years: Should we provide people with a terminal illness such as AIDS access to experimental, unapproved drugs? Is it ethically acceptable to prevent people with life-threatening illnesses from accessing experimental drugs in order to motivate them to join clinical trials where such drugs might be tested?

Since I wrote this book, combinations of nucleoside analogues and a new class of drugs, the protease inhibitors, have been demonstrated to substantially prolong the life expectancy of HIV-positive people. Insufficient time has passed to answer the question of whether these combination therapies are capable of preventing HIV from developing resistance to these drugs. However, more recent data show the impact of protease inhibitors may not be as long-lived as originally thought.

In any case, the ethical questions discussed in this book have relevance beyond the case of AIDS, which was chosen as a real-world case study. Indeed, the answers proposed and defended in this book have implications for many people suffering from a variety of terminal illnesses. It is my hope that patient-lobby organizations will have a closer look at what AIDS activism has achieved over the last decade and how it was achieved. I believe that the application of this knowledge might help to improve the situation of people with terminal illnesses who are currently pressured into joining clinical trials for the purpose of testing experimental drugs. As a corollary of increasing volunteerism in terms of trial participation, the quality of clinical research would also be improved.

Research for this book was undertaken between March 1992 and the end of 1995, when I was a PhD student at the Monash University Centre for Human Bioethics in Melbourne, Australia. Peter Singer supervised the original research. Without his ingenious criticism, the content of this book would look quite different. Dieter Birnbacher and David Seedhouse read parts of earlier drafts of the manuscript and provided me with ingenious comments that subsequently led to a number of revisions. My work was supported by grants provided by the German Academic Exchange Service and by Monash University. I am grateful to both institutions for their support. In England, the Centre for Professional Ethics at the University of Central Lancashire provided funding to translate my "funny prose" into publishable text. Jacinta Kerin undertook this editorial work. Louise Williams, the Centre's secretary, corrected a large number of errors created by faulty software. Last, but not least, I should mention Carlton Hogan, editor of *PWALive* in Minnesota, who taught me a great deal about clinical trial designs. I thank him and all those mentioned above for supporting this work.

Introduction

The objective of this book is to answer the following question: Are there good ethical reasons to prevent people with terminal illnesses such as AIDS from buying and using experimental (unapproved) drugs? In addressing this question, I will first analyze paternalistic reasons that might justify such prevention. I shall then consider arguments alleging that to allow access to experimental agents constitutes harm to others. In the context of the latter, I will raise the question of whether the current clinical trials system requires ethically unacceptable and/or practically impossible high levels of altruism from the prospective patients joining these trials.

The majority of this book, therefore, explores ethical issues in regard to access to experimental drugs for people with terminal illnesses and ethical issues related to the design of research clinical trials. Finally, in the case that the argument for allowing such patients to use experimental drugs is accepted, I will attempt to answer the question of who should pay for access to experimental agents for people with terminal illnesses.

Many people suffer from a variety of terminal illnesses, the most notorious among them being, perhaps, certain types of cancer and, of course, AIDS. No successful standard treatments for these diseases that could properly be called therapy have yet been developed. Palliative care may be available, or as in the case of AIDS, prophylactics and therapies for some opportunistic infections are available, but ultimately, people suffering from these diseases will succumb to them. Of course, many researchers work very hard to develop therapies for such terminal illnesses.

To achieve this goal, they run laboratory experiments and, at a later stage, preclinical and clinical trials. Should they succeed, patients will be glad to have a working drug available. Until then, however, primary care physicians can only do their best to keep these patients alive, attempting to preserve their quality of life for as long as possible, at levels as high as possible. In the past, patients have more or less accepted their individual fates, and have not even considered acting against the very medical system that is in place to keep them alive. They have not considered it wise to take on medical researchers and standard clinical trial procedures in last-ditch attempts to save their lives.

AIDS, however, has changed this picture drastically. Highly sophisticated and creative activist groups such as ACT UP have not only raised public awareness concerning AIDS with immortal slogans such as SILENCE = DEATH (on the background of a pink triangle signifying the experiences of gay men in Nazi concentration camps),[1] "die-ins," and other such publicity-gaining symbolic acts, they have also vehemently criticized the recalcitrance of the established research agencies to respond to the crisis.[2] They have charged the government with a number of different crimes against their survival interests. One such charge was that governments made research funds available too little, too late. Those communities that were hardest hit by AIDS were certainly entitled to make this charge. As long as it was believed that AIDS was nothing other than a gay-related immunodeficiency, (GRID, as it was called in those days),[3] both the public interest and the flow of research funds were negligible.[4]

However, once these funds did become available, as they are today in the case of AIDS, the activists' targets changed. They then questioned the status quo in two different, but related, areas of the testing and availability of drugs. They argued that, given the slowness of the AIDS research establishment's drug testing and approval process, they should have access to experimental drugs. Furthermore, they criticized the inability of the government to enroll all patients who wanted to participate in clinical trials.[5] Ultimately, their point has been that, in cases of terminal illness, no government has a moral right to prevent patients from taking their chances with experimental agents of their choice.

Such activist groups have also criticized the standard procedure of giving placebos in research clinical trials as inappropriate for patients suffering from terminal illnesses. Joseph Sonnabend, a physician who was the medical director of the New York City Community Research Initiative on AIDS, described this problem:

One of the greatest difficulties in the conduct of AIDS treatment trials has been that the effects of a potential intervention had had to be measured against a higher rate of disease progression or more deaths in a group receiving a placebo compared to one receiving an active substance. In other words, *individuals may be required to die in order to demonstrate the efficacy of a particular treatment.* Since there is no treatment for AIDS, participation in a trial may represent the individual's only hope of receiving the *possibility* of treatment. This highlights the ethical concerns that make the use of placebos unacceptable to people whose life expectancy is of a similar order as the duration of the trial.[6]

However, it was not maliciousness that led governmental authorities in most countries to institute a drug approval mechanism that prevents drug producers from selling these agents, without governmental approval, to the public. In part, it has been negative historical experiences that have led to the regulation of the drug approval process by governmental agencies.

Similarly, good historical reasons exist for introducing placebos into the clinical trials process. An example of the former can be found in Ellen J. Flannery's (1986) report of North American "quacks" successfully selling huge numbers of boxes in the 1930s. The salespeople claimed (and promised) that these boxes would heal all known diseases. Of course, the boxes did not heal any diseases, and thousands of people wasted their money on the purchase of this device.

The widespread acceptance of the view that the state bears some responsibility for organizing health care for its citizens has led to the creation of numerous institutions which are thought to support the state in fulfilling its obligation to its members. A number of administrative and regulative bodies have the function of controlling the safety and efficacy of new drugs that pharmaceutical companies would like to market. Their primary (but by no means only) function is the protection of consumers' rights. For instance, the U.S. Congress accorded the Food and Drug Administration wide-ranging powers, which changed the system of premarket notification, (in place at that time) to a system of premarket approval. "After 1962 the FDA had affirmatively to say yes. . . . Congress also required the FDA to evaluate drugs not only for safety [phase I trials] but for efficacy [phase II and III trials] as well."[7]

The FDA is now legally obligated to approve the general marketing of drugs only if no (or only negligible) doubts remain regarding the safety and efficacy of an experimental drug for the prospective users. Edgar and Rothman are certainly justified in their conclusion that "it is not surprising that the FDA defined its goal after 1962 in terms of minimizing risk."[8] The rationale behind the establishment of such governmental institutions, and their possession of wide-ranging regulatory powers, can be found in drug scandals, such as elixir sulfanilamide (1937), diethylstilbestrol (1950-1952), and thalidomide (1962), which led to enormous suffering in the population. Unscrupulous drug companies, supported by scientists acting in an equally unethical manner, have repeatedly offered highly toxic drugs that had not undergone preclinical and clinical trials to large parts of the community; their safety and efficacy had not been verified in standardized, reproducible clinical trials. The current status quo, however, has come under fire from the new conservative majority in the U.S. Congress.[9] The Republican majority's drive to deregulate the American federal bureaucracy has also focused on the FDA.

In the case of AIDS research in America, it is the responsibility of the FDA's Anti-Viral Drug Advisory Committee to advise the FDA regarding applications of pharmaceutical companies to either approve a new drug or to change the indications upon which it is possible for a physician to prescribe it. To do this, the committee must answer this question: has safety and efficacy been scientifically demonstrated by adequate and well-controlled trials? The scientist Alvin Novick, who has joined the committee as a consumer representative, described some of the shortcomings in the work of this institution that are due to its limited resources and executive powers:

> A window on the world of drug development and regulations opens for the committee members three or four times a year for a period of one to two weeks each time. The window has mechanical guards that strictly regulate how widely it can be

opened. Much is kept private. The sponsoring company for a new drug shares what it wishes. . . . The committee's power to expand the opening or to clarify its transparency is limited to raising questions at meetings. Answers or full answers need not always be given or may be given in a language that is evasive or challenging.[10]

Today the development of new drugs is a tedious, lengthy process that requires a significant number of preclinical and clinical trials before final market approval or rejection is given in regard to a particular agent. It is not uncommon that several years, sometimes more than a decade, may pass between a given drug's first appearance on a drawing board in the research laboratories of a pharmaceutical company and its final approval by the responsible authorities. This process is not only time-consuming, it is also extremely expensive for the pharmaceutical company that intends to apply for the approval of a new substance. The company must finance the whole research process in advance.

Not surprisingly, the monopolization and building of huge pharmaceutical trusts continues. In the end, of course, it is the consumer who pays the price for the high safety measures, which have been installed to protect his or her interests. The reasons for measures such as these are obvious: ill people should be protected from financial exploitation by quacks or from serious harm by drugs that either do not work or that have severe side effects.

Another reason for the tight rules in the drug approval process is the belief that independent (governmental) control of clinical research generally leads to reliable results that will be of great significance for the well-being of a great number of prospective patients. Thus, they have, at least prima facie, reason to think that the drug their physician has prescribed to them will really work.

This drug approval process is the same, regardless of whether the drug in question is merely thought to help persons with a cold or is supposed to combat terminal cancer. For patients with terminal diseases (where by definition no successful standard therapy does exist), no life-saving drugs are available. The current laws in most countries prevent such persons from buying and using unapproved drugs at their own risk.[11] Prima facie it looks as if the autonomy of persons with terminal diseases is violated in this situation because they are denied the opportunity to choose alternative courses of treatment for a disease that is likely to kill them in the near future. So how is this status quo justified? One of the reasons given refers to the "interests of the whole population," usually cloaked in the language of "public health," and specifically, to the health of prospective patient-generations.[12]

The moral argument for these two sets of interests runs along the following lines: it is impossible to find out whether an experimental agent is safe and efficient without randomized, placebo-controlled, double-blind clinical trials. Current patients, when presented with the alternative to join such a trial or to buy and use the agent in which they are interested (without a 50 percent chance of getting a placebo or another agent they are not keen to use), would not join the trial. Hence, the argument is that it is necessary to close all other avenues that might allow those patients access to these drugs.

A different kind of argument for denying terminally ill people access to experimental drugs is paternalistic in nature. It does not, as does the preceding one, rely upon a strategy asserting that we have moral obligations to future generations to act altruistically (to a degree that might force us to sacrifice our lives). Rather, this argument claims that we cannot allow people with AIDS to buy and use experimental drugs because they could harm themselves to a degree that is unknown. This harm could range from insignificant and reversible damages to irreversible damages that may have a severe negative impact on the quality of life that even people with AIDS still enjoy.

Since the risk for patients who join controlled clinical trials is significantly lower (because they are under permanent clinical observation), it is argued that these trials should remain the only means by which experimental drugs can be accessed. One corollary of this argument is not paternalistic; it argues that such persons who have harmed themselves through experimental drug use might become permanently dependent on social welfare. It is claimed that this dependence might last much longer than that of the average person with AIDS who is not using unapproved drugs. Hence, their behavior harms others. I will discuss this argument in Chapter 3 of this book.

People with AIDS, however, have made strong accusations, insofar as the design of clinical AIDS trials is concerned. They claim that, more often than not, there is a primacy of research interests over the survival interests of the participating patients.[13] Chapter 2 of this book will discuss the question of whether this criticism is justified: are there clinical AIDS research studies in which the researchers considered the survival interest of their research subjects to be of less importance than their immediate research interests?

If such studies exist (as I will show they do), then the inevitable ethical question is, are there good ethical reasons that might justify studies with such a design? I will also address the defensibility of the suggested ethical justification for the expectation that people with terminal illnesses who join clinical trials should accept major disadvantages for their own survival over the course of the trial. In this context, I will discuss the problem of whether the participation of terminally ill patients in clinical trials can reasonably be called voluntary, given that they generally have no access to experimental agents outside of a clinical trial.

AIDS communities in the United States, which basically consist of gay men and intravenous drug users (IDUs), found support from quarters of the American mainstream that are not normally considered to be the natural political allies of the activists from these subpopulations.[14] *The Wall Street Journal,* for example, happily embraced their political goals. After all, this was a good chance to continue the deregulation policies introduced and implemented by former U.S. Presidents Ronald Reagan and George Bush. David Henninger, for instance, asked innocently, "By what legal or moral right do we abide a system that tells huge numbers of gravely ill Americans they cannot try these [experimental] therapies until a bunch of people in a federal building in Rockville, Maryland, say so?"[15]

The first part of Henniger's rhetorical question can easily be answered by those federal U.S. laws that give the FDA the legal right to prevent terminally ill people from trying experimental therapies. The question that remains to be addressed, however, asks whether or not the state has the moral right to prevent dying people from trying experimental agents in a last-minute attempt to save their lives.

The structure of this analysis will be as follows: Chapter 1 discusses the centrality and importance of individual autonomy and self-determination in our lives. I will analyze several classical, as well as some contemporary, concepts of autonomy and discuss the question of where these theories leave paternalistically motivated intervention strategies. This chapter demonstrates that neither the currently predominant theories of autonomy (be they deontologically or teleologically motivated) nor their correlative theories regarding weak and strong types of paternalistic interventions are sufficient to ethically justify the current situation, which prevents terminally ill people from buying and using experimental drugs.

In Chapter 2, I shall analyze certain other arguments which support the view that people with terminal illnesses should not be given access to experimental agents. These arguments rely on the idea that, if we were unable to motivate these patients to join clinical trials (an alleged consequence of access to experimental agents), we would be unable to conduct adequately designed trials, and hence a much greater number of prospective future patients would be harmed because drugs could no longer be developed and tested. I will argue that we have no empirical evidence for the claim that useful drug trials would no longer be possible.

Furthermore, I will demonstrate that people with AIDS are often expected to sacrifice their individual survival interests in badly designed research clinical trials. A direct consequence is that many of these patients begin to "cheat"; that is, they do not report concomitant medications while remaining on the trial medication in order to access the experimental drug they wish to use. As a result of this occurrence, such trials lose most of their predictive power. I suggest that the expectations of altruistic behavior of terminally ill people are far too high and unrealistic. I propose instead to allow people with AIDS (and other terminal illnesses) to access experimental agents and to conduct clinical trials only with true volunteers (as opposed to the current situation in which people join clinical trials because they are denied access to experimental agents through other means). I also suggest a few practical measures for using the information we can gather under such circumstances in order to further our knowledge about experimental agents.

Finally, in Chapter 3, I address practical issues, such as who should pay for the experimental agents and who should pay should the risk taking of the terminally ill patient lead to harmful consequences for the patient.

Chapter 1

Autonomy—Access to Experimental Drugs and the Terminally Ill

In the Western industrialized nations, as well as in the democracies of the former Eastern Bloc, individual liberty rights have a fundamental importance. They constitute the ideological basis of pluralistic, secular societies. They rely on the central premise that the state, as well as the law, functions primarily to prevent harm to others. Exceptions are generally made in instances where the acting individuals are considered incompetent and, hence, unable to act autonomously. In these cases, in the opinion of most moral philosophers (independent of the moral or ethical theory to which they subscribe) with whom I tend to agree, it is ethically justifiable to intervene in individual lives in order to prevent self-harm.

Limitations of this individual right to self-determination, however, do not exist only in regard to people who are considered incompetent by other parties. For instance, adult homosexuals are still discriminated against even in many democratic, liberal societies for sexual acts that heterosexuals are permitted to commit without any negative legal consequences. It should be noted, however, that unequal, discriminatory treatments on the basis of sex are being progressively rejected in Western countries through the successful campaigns of civil rights organizations such as the American Civil Liberties Union or the Humanistische Union (Humanist Society) in Germany.[1] This reflects a general tendency that the majority of people no longer approve of discrimination on the basis of sex.

Another example of the increasing recognition of an individual's right to self-determination is provided by the example of legislation regarding suicide in Australia. Until recently, suicide was illegal in most Australian states. However, the insight that it is quite absurd to penalize someone for attempting to not only live one's own life by one's own values, but also to end that life in accordance with personal ideas about how to die with dignity, has led to the legalization of suicide in all states of the Australian Commonwealth, with the notable exception of the Northern Territory, where suicide remains illegal.[2]

The attitude of the majority of states in Australia seems to be more in touch with the ordinary citizens' opinions on this subject matter, in comparison to the Northern Territory's legislation. It is naive to believe that it is viable to penalize people who have decided that their life is no longer worth living. Clearly, a person who has decided to commit suicide will not be very impressed upon learning that any such attempt is illegal and might lead to prosecution. It is also quite unrealistic to believe that such a measure can deter people from attempting suicide in the first place.

Respect for the individual autonomy of its citizens is a central, perhaps fundamental, element of Western democratic society. This respect finds its expression not only in the already mentioned national constitutions, but also in international declarations, conventions, and contracts of the United Nations. This underlines the idea that such respect is not only fundamental, but also universally approved. Its scope reaches beyond capitalistic-democratic Western ideology because it has been endorsed by all member states of the United Nations, independent of the geographical location of the country and of its current political (or religious) system.

Autonomy is a *terminus technicus* that goes back to the Greek language. It is derived from the Greek words *autos* (self), and *nomos* (rule). Historically, it was probably first used to describe the self-ruled attic Greek. Citizens had the democratic power to create their own legislation instead of having it forced upon them by external despots.[3] Respect for individual autonomy signifies an extension of the self-rule of a democratic society to self-rule of the citizens in such a society.

Self-determination of an autonomous person means, ideally, that this person is not subjected to any kinds of controlling external or internal constraints. The autonomously acting individual must, as Robert Young puts it quite succinctly, "freely direct and govern the course of his [or her] own life."[4] The interpretation of the term autonomy is (perhaps inevitably) not uniform within philosophical literature. We are confronted with a number of different accounts of autonomy, since diverse philosophical viewpoints have led to equally diverse definitions of autonomy.

On a broader scale, this situation has led various Western democracies to interpret autonomy differently than other democratic countries, resulting in contrasting interpretations that have engendered different practical conclusions as to which individual liberties people should have in their attempts to live their lives. Moreover, the questions that are widely discussed are not necessarily as contentious as are the claims, for example, of homosexual people to have the right to satisfy their individual sexual interests.[5]

The problems are sometimes seemingly of much greater triviality and comprise issues such as the appropriateness of making helmets compulsory for cyclists or, indeed, whether it is ethically acceptable to permit smoking in public places.

In the area of AIDS, a classical contentious issue is the alleged conflict between public and private interests. An example of this might be found in the persistent debate over whether to close gay bathhouses. The argument between proponents of the idea to close gay bathhouses in order to reduce the risk of HIV infection for gay men (interpreted as harm to others)[6] and those who argue that they should remain open (because HIV infection in such settings can be interpreted as constituting harm to self)[7] is still raging, even in AIDS activist circles.

John Stuart Mill's interpretation of autonomy, as articulated in his classic work *On Liberty*, is basically concerned with freedom from coercion.[8] If one subscribed to this reading, then it might be immediately obvious that our autonomy has become significantly restricted through paternalistic governmental rules, such as those that require everyone who rides in a car to wear a seat belt. According to Mill, human beings are reduced to machinelike entities when deprived of the freedom to live their own lives and to make choices in that regard. He believes such humans lose the ability to express fundamental human characteristics:

> He who lets the world, or his own portion of it, choose his plan of life for him, has no need of any other faculty than the

ape-like one of imitation. He who chooses his plan employs all his faculties. He must choose observation to see, reasoning and judgment to foresee, activity to gather materials for decisions, discrimination to decide, and when he has decided, firmness and self-control to hold his deliberate decision. . . . It is possible that he might be guided in some good path, and kept out of harm's way, without any of these things. But what will be his comparative worth as a human being?[9]

Thus, according to Mill's metaphor, humans are reduced to mere machines. A significant curtailment of their freedom would take from them exactly what signifies human dignity, that is, the capability to act "according to the tendency of the inward forces which make it a living thing."[10] I still consider it problematic that the current situation forces autonomous beings to wear seat belts or face the prospect of getting fined. In my opinion, it seems prudent to use a seat belt to reduce the likelihood of becoming severely injured in case of a car accident. It also seems prudent to try to convince citizens of this view and therefore to organize information campaigns to succeed in that goal.[11]

For the vast majority of people living with AIDS, it is probably more important than for most other citizens that their individual interests to direct the course of their own lives are respected by society.[11] It is likely that few (if any) competing values will have similar importance in the lives of these persons. The reason for this is not, as one might initially be tempted to think, that they have a terminal disease. Such a reason is not AIDS-specific, since many people die of other terminal diseases.

Another, more compelling reason exists to suspect that respect for individual autonomy is of greater importance for people with AIDS than for persons with other lethal diseases: most people with AIDS in Western industrialized nations,[12,13] as well as in a number of Southeast Asian countries,[14,15] belong to classically marginalized groups of our societies. They are either homo- or bisexual men, injecting drug users (IDUs), or both. These risk groups constitute more than 90 percent of all AIDS cases in Western societies.

Such people with AIDS share important life experiences that they do not have in common with most other citizens of our societies. Homo- and bisexual men lived, until very recently, lifestyles that were deemed illegal. In Australia, for example, homosexual sex acts were illegal and prosecuted, even when they occurred between consenting adults. Many of these men have now organized themselves in the modern gay liberation movement.

Homosexual behavior is considered illegal in many other countries. For instance, the Singapore penal code threatens sentences ranging from two years to life imprisonment for those engaging in homosexual acts. Singapore has gained worldwide notoriety with undercover police actions, including trying to seduce gay men into illegal behavior. Such police action is sanctioned by this country's government-controlled media.

Homosexual relationships generally remain unacknowledged by most national legislations. They do not, for instance, exist in taxation laws. Most homosexual men, at some stage of their lives, have had to decide whether to abide by a law they considered immoral, or to live an autonomous life at the cost of breaking this law. Thus, many homosexual men view with suspicion attempts to direct their lives through governmental regulations. This is even more true for medical researchers. After all, professional psychiatry's century-old view of homosexuality as a mental disease has not been forgotten by many of these men.[16,17]

The situation of IDUs is comparable to that of homosexual men.[18] They are not only targets of societal anger, but also targets of police persecution. Their respective attempts to legalize soft and certain hard drugs have failed in virtually every country of the Western hemisphere. IDUs are the modern outlaws in our societies. The current "war against drugs," which was introduced by former U.S. President Ronald Reagan, is primarily a war against the users of illegal drugs. The decision of many users of such drugs to violate drug laws has cost a great many of them their freedom, if not their lives. Evidently, IDUs have more in common with homo- and bisexual men, as far as their suspicion of governmental or medical authority is concerned, than with white middle-class citizens suffering terminal cancer. The only difference relevant to this discussion seems to be that homosexual behavior is no longer illegal in a number of countries. However, the annual statistics regarding "gay bashings" indicate strongly that there is a significant difference between the legalization of homosexual behavior and the societal acceptance of homosexuals.[19,20]

I agree with Robert Young, who laments that it is "surprisingly difficult" to describe exactly what makes many of us value "personal autonomy as a character ideal."[21] However, the behavior of homosexual men or IDUs leads me to the opinion that it is not so much the intrinsic but rather the instrumental value of individual autonomy that motivates us to hold it in such high esteem. Individual autonomy is so important because it allows us to develop convictions and value systems, to reflect upon these values, and to change or retain them, just as *we* deem it appropriate. It also allows us to develop a life plan, if we wish, and to decide whether or not we will ultimately try to live consistently with the values and goals we have set out in this plan.

Individual autonomy also permits us to strike a balance between current interests and long-term values. Autonomous individuals alone are able to make a reflective judgment about whether they should violate their medium- and long-term goals in order to satisfy actual interests. To act autonomously means that an individual is able to try, in reality, to follow up his or her interests and to satisfy them.

Theoretically, respect for individual autonomy has great importance in all Western democracies. Practically, however, the governments of most countries often violate individual autonomy rights. Perhaps one reason for this is that it is unclear how far-reaching the rights of an individual to live his or her own life are. These rights are often in conflict with the teachings of the major religions, which has led and still leads to intrusion into the lives of individuals who exercise their right of self-determination.[22] Other, more fundamental and complex reasons are to be found in the fact that there is still no consensus regarding the question of what autonomy is and which necessary or sufficient conditions must be met before a certain act can be considered an autonomous act. It also remains unclear as to which ethical reasons can be used to justify intrusions in the lives of autonomous individuals. Is it only allowed when they harm others? Or are we entitled to prevent these persons from actions we consider harmful to themselves?

In this chapter, I will discuss a number of concepts of individual autonomy in order to answer the first question set out in the introduction: Is it possible to use paternalistic arguments to provide an ethical justification for preventing persons with AIDS from buying and using unapproved, experimental drugs? John Kleinig has described this variation of paternalistic interventions as "passive paternalism" because it tries to prevent persons from doing things they want to do on the basis that they may harm themselves and that such action cannot be in their best interest.[23]

Thus, the structure of the rest of this chapter will consist of evaluations of classical, as well as modern, ethical theories in regard to whether they would support paternalistic interventions that have the goal of preventing the use of experimental drugs in the lives of people with terminal illnesses such as AIDS. The first major author to be discussed is Immanuel Kant.

Admittedly, the paragraphs analyzing Kantian ethical theory sit uneasily in the flow of my argument. This is largely due to the fact that Kant himself says little in regard to whether people with terminal illnesses should be allowed access to experimental therapies. Nevertheless, I have a number of reasons for discussing Kant's work, namely his significance as perhaps the most important philosopher in German history and his considerable influence on philosophers such as Dworkin and Young, whose ideas I will discuss later in this chapter. Another reason I start with Kant is historical: his works are the oldest among those that I have chosen to analyze more closely.

IMMANUEL KANT

As the German philosopher of the enlightenment, Immanuel Kant's (1724-1804) major concern is the basic problem of the freedom of the will, discussed in his *Groundwork of the Metaphysics of Morals*. This has, of course, relevance to the questions that are at the core of this book, insofar as autonomy is concerned. Kant distinguishes between *officia iuris* and *officia virtutis*. The former implies an obligation to the positive law as such, while the latter implies a moral duty to the moral principle which, Kant claims, does exist.[24] The moral value of my action lies, in Kant's opinion, not in the intention I have, but in the reason I present as to why I have done a certain thing.[25]

He writes in the *Critique of Practical Reason,* "The disposition which obliges him to obey it is: to obey it from duty and not from spontaneous inclination or from an endeavor unbidden but gladly undertaken."[26] Kant's theory starts, just as much as Mill's (see next section), from the assumption that beings who have a will and are capable of wanting something have the capacity to choose (if there is a procedural independence) and that, therefore, these individuals will inevitably encounter situations in which they must make choices.

From where should we derive the moral law that is to guide their acts? Is it really possible, as Kant claims, to discover an objective, noncontingent moral law that possesses universal validity and is independent of all the laws of causality of the physical world?

Immanuel Kant's strategy was consistent with his successful distinction between pure (a priori) knowledge and empirical (*a posteriori*) knowledge in his theoretical philosophy. He believed that such an a priori identifiable law does indeed exist. It is "not in the nature of man, nor in the circumstances of the world in which he is placed, but solely *a priori* in the concepts of pure reason."[27] He warns that we should not assume, or even attempt to "derive the reality of this principle from *special characteristics of human nature.*"[28]

Thus, before we have even arrived at Kant's supreme principle of the doctrine of morality something is already clear: if this principle is, as Kant claims, a priori evident, then it must be a synthetic statement. It has to be a formal principle that is devoid of any content of acts undertaken in the empirical world. That this principle could possibly be the first, supreme principle of morality inevitably evokes skepticism. Kant's devoted admirer, Arthur Schopenhauer (1788-1860), can barely hide his doubts: "Let us consider what this means," he writes, "human consciousness as well as the whole of the external world together with all experience and facts in them are swept away under our feet. We have nothing on which to stand. But what are we to hold onto?"[29]

It is also apparent which conditions a being has to meet before a Kantian can accept him or her as a moral agent. It is not sufficient for this entity to act out of mere interest or desire because this would not meet the requirement to act out of duty. As Kant writes, "Everything in nature works in accordance with laws. Only a rational being [a necessary condition to become a moral agent] has the power to act *in accordance with his idea* of laws—that is in accordance with principles—and only so he has a will."[30] Schopenhauer considered this reason to constitute a *petitio principii* because he thought that we have no reason to assume that there are laws to which our conduct ought to be liable.

However, Kant thought that only a being who is capable of acting this way has the potential to act autonomously. All other living beings (such as nonhuman animals) act, in Kant's view, on the basis of heteronomous principles. Hence, a patient's desire to survive or to avoid pain is unacceptable as a motive for a morally good action. Such interests are heteronomous, and an action that leads, in consequence, to the avoidance of harm is, as such, not good. That is, it has no moral worth.

However, it may still be the case that the action itself is in accordance with what we are actually required to do. In this case, the action and its consequences can still be considered right. The action can become a morally good one if the patient's motive was, ironically, not primarily to survive or to avoid harm, but to follow a general moral principle that would require such an action. People with AIDS who would like to use unapproved drugs would, therefore, have to convince a Kantian that their motive for wishing to buy and use unapproved experimental agents is *not* based on their interest in surviving but is based solely in a general moral principle upon which they act. Thus, under a Kantian schema, any patients' motives that were based on their own survival interests would have no moral worth. Not only is this idea theoretically questionable, it is also of dubious practical relevance, since it seems highly unlikely that many terminally ill patients could really be motivated in the way in which Kant deems appropriate.

Kant's answer to the question of the supreme ethical principle has become famous in the formulation of the Categorical Imperative: "I ought never to act except in such a way *that I can also will that my maxim should become a universal law.*"[31] He has presented this same idea in slightly varied explanations in his *Groundwork*. "There is therefore only a single categorical imperative and it is this: *Act only on that maxim through which you can at the same time will that it should become a universal law.*"[32] The Practical Imperative that is supposed to deliver the guiding principle of our acts is this: "Act in such a way that you always treat humanity, whether in your own person or in the person of any other, never simply as a means, but always at the same time as an end."[33]

Kant never tires of stressing that the justification of this supreme principle of morality cannot be found in empirical evidence proving its utility. It is also impossible, in his view, to prove that it flows somehow from our moral intuitions, because "an action done from duty has its moral worth *not in the purpose* to be attained by it, but in the maxim in accordance with which it is decided upon."[34] The Kantian argument claims here that the form (the universality) of his supreme principle is at the same time its justification. His argumentative strategy becomes most apparent in the *Critique of Practical Reason:*

> The mere form of a law, which limits its material, must be a condition for introducing this material into the will, but not for presupposing it as the condition of the will. . . . If I attribute this to everyone . . ., it can become an objective practical law only if I include within it the happiness of others. Therefore, the law that we should further the happiness of others arises not from the assumption that this law is an object of everyone's choice but merely from the fact that the form of universality, which reason requires as condition for giving to the maxim of self-love the objective validity of law, is itself the determining ground of the will. Therefore not the object, i.e., the happiness of others, was the determining ground of the pure will but rather it was the lawful form alone. Through it I restricted my maxim, founded on inclination, by giving it the universality of a law, thus making it conformable to the pure practical reason. From this limitation alone, and not from the addition of an external incentive, the concept of obligation arises to extend the maxim of self-love also to the happiness of others.[35]

This explains why Kant cannot derive value judgments regarding acts from these acts' consequences. After all, he would argue, such consequences can be consequences of other causes too. There is no need for the will of an autonomously acting, rational being. However, in this "will of a rational being, . . . the highest and unconditional good alone can be found."[36] As the preceding quote indicates, it seems that Kant thinks that the will of an autonomous being is inherently valuable.[37] He must also hold, however, that it is of instrumental value because this will is necessary to analyze whether our options to act would do justice to the Categorical Imperative in meeting its conditions. Only moral persons can overcome this Kantian hurdle. "A *Person* is a subject whose actions can be *imputed* to him." Only such persons are fully autonomous in their actions. "*Moral* personality is therefore nothing other than the freedom of a rational being under moral laws. . . . From this it follows that a person is subject to no other laws than those he gives to himself (either alone or at least along with others)."[38]

In light of Kant's belief that a value judgment about the morality of a certain act cannot be derived from the act's consequences but only from an answer to the question of whether a person has acted out of duty to the moral law, it has commonly been argued that Utilitarianism and Kantianism are incompatible. This has been held to be the case because the first is teleologically oriented while the latter is a deontological theory. It is important to keep in mind, however, that the Categorical Imperative claims only to be the supreme principle of morality. It tells us nothing about the content of moral principles or practical ethical norms. The Categorical Imperative would only demand that these principles and norms have a certain (universal) form.

Thus, the conclusions of Kant's philosophy of law, as they are laid out primarily in his *Metaphysics of Morals,* are, in part, compatible with Mill's views of the limits of individual autonomy, as far as the interests of persons other than the acting individual are concerned. This is true, for instance, in the context of Kant's ideas of the limits of individual liberty. He wrote, "If then my action or my condition generally can coexist with the freedom of everyone in accordance with a universal law, whoever hinders me in it does me *wrong*."[39]

Clearly this passage is incompatible with John Stuart Mill's philosophy in at least one vital respect, which is in regard to acts whose consequences concern only the acting individual. Mill's individual could not care less whether his or her acts are in "accordance with a universal law." The situation is different, however, when it involves actions that affect others. As we will see later, Mill himself believed that his views were compatible with Kantian philosophy. He argued:

> [Kant] virtually acknowledges that the interest of mankind collectively, or at least mankind indiscriminately, must be in the mind of the agent when conscientiously deciding upon the morality of an act. . . . To give any meaning to Kant's principle, the sense put upon it must be, that we ought to shape our conduct by a rule which all rational beings might adopt *with benefit to their collective interest.*[40]

Richard M. Hare supports this interpretation in a recent paper, arguing that Kant's second expression of his supreme principle, the Practical Imperative,[41] "can certainly be interpreted in a way that allows him—perhaps even requires him—to be one kind of utilitarian."[42] Indeed, Hare presents the reader with a number of quotes from Kantian writings that allow such an interpretation. For instance, to meet the requirement of universality of the Categorical Imperative, the moral actor (who must, following Kant, be treated as an end in himself) makes others' ends his own (unless they are immoral).[43] Hare argues that Kant could have been some sort of rational-will utilitarian, insofar as actions affecting others are concerned because "a utilitarian too can prescribe that we should do what will conduce to satisfying people's rational preferences or wills-for-end—ends of which happiness is the sum."[44] J. J. C. Smart, who proposed a Kantian type of rule utilitarianism, has made a similar point earlier.[45]

Unfortunately, Kant asks us to treat autonomously acting individuals, as well as ourselves, always as ends in themselves or ourselves, respectively, without having explained anywhere in his works what it could possibly mean for anyone to be an end in themselves. H. J. Paton's explanation, which is to describe rational actors and persons as ends in themselves and to claim that they alone have "unconditioned and absolute value," does not really help to solve this mystery.[46] After all, as Schopenhauer argues persuasively, to be an end is to be the direct motive of an act of the will. He criticizes Kant's postulation of humans (and also of every other imaginable rational being) as existing as ends in themselves, when he concludes his analysis:

> I must say frankly that "to exist as an end in oneself" is an unthinkable expression, a *contradictio in adjecto*. . . . Nor does the absolute worth that is said to attach to such an alleged but inconceivable *end in itself* fare any better, for this too I must without mercy stamp as a *contradictio in adjecto*. Every *worth* is a quantity of comparison and even stands necessarily in a double relation. First, it is relative, in that it exists for someone; and secondly, it is comparative, in that it exists in comparison with something else by which it is valued or assessed. Outside these two relations, the concept worth loses all meaning; this is so clear that there is no need for further discussion.[47]

Indeed. Nevertheless, the notion of a human being as an end in itself is of crucial importance for Kant's valuing of individual autonomy. Respect for individual autonomy, which was one of the central features of the enlightenment, follows from the unconditional worth he has ascribed to each person. If I do not respect that autonomy, I treat it merely as a means.

As far as self-regarding actions are concerned, Kant requests that we treat ourselves (and humankind represented in us) never as a means and always as an end. It is only coherent with this view that our power of disposition over ourselves finds its limits where we would begin treating ourselves as mere means to certain ends. To reiterate, however, despite all the rhetoric surrounding this idea, it remains unclear what Kant could possibly have meant by the concept of rational beings as ends in themselves.

For instance, when I use my body in order to cycle to the university every day, I clearly use myself as a means to a certain end, which is to reach the university. It is not clear, however, what Kant could have meant when he derives a general prohibition of such uses of ourselves as a mere means, as opposed to an interpretation of ourselves as ends. His practical conclusions are surprisingly consistent with the teachings of Christian moral philosophy. For instance, he condemns suicide and homosexual acts in the strongest possible terms.[48] The reason for condemning the former is given in his *Lectures on Ethics:*

> We may treat our body as we please, provided our motives are those of self-preservation. . . . In taking his life he does not

preserve his person; he disposes of his person and not of his attendant circumstances; he robs himself of his person. This is contrary to the highest duty we have towards ourselves, for it annuls the condition of all other duties; it goes beyond the limits of the use of the free will, for this use is possible only through the existence of the Subject.[49]

This alleged self-abuse leads Kant to the interpretation that whoever commits suicide degrades himself or herself to a mere means, because he or she "disposes of his person," and this is only appropriate in the case of things but is never acceptable when persons are concerned. Therefore, it is impossible to morally condone any suicide or to ever approve of it because a person who commits suicide uses his or her body as a mere means to overcome suffering.

Practically, Kant's ethical theory would lead to a number of real-world absurdities. For example, the same act by two people with AIDS might lead to praise of one person's act, but a verdict of "no moral worth" for the other's, depending upon the motive they present as a justification for their course of action. One might be heavily criticized for using unapproved drugs because it was done in order to save the person's life or to reduce suffering. The other might be praised for the very same course of action because it was done out of duty to a still-to-be-discovered general ethical principle that requires the person to do just what was done.

The great differences between Kantian and Millian approaches to alleged duties to ourselves become obvious when we consider Kant's view that suffering (no matter how great it might be) is not a sufficient reason to commit suicide. He answers his rhetorical question, "Can I then take my own life because I cannot live it happily?", with an unequivocal "No!" "All our duties towards ourselves would then be directed towards pleasure."[50] Assume that a devoted Kantian suffers from a disease that comes with severe pain, and he or she needs to decide whether to choose a drug that reduces, or even eliminates, the pain at the cost of a shorter overall life span on the one hand, or no drug, severe pain, and an overall longer life span on the other. Kant would never accept the decision to take the drug because "misery gives no right to any man to take his own life."[51]

It seems to me that Hare's view of Kant's claimed duties against ourselves is not too far away from the truth, when he suspects that Kant "perhaps is harking here back to something he heard when young, that man is created as a human being to fulfill an end ordained by God."[52] This interpretation follows closely along the path of Schopenhauer's critique of the Kantian ethic. He concluded in 1840, "thus even Kant's basis of ethics, considered for the last sixty years as a firm foundation, sinks before our eyes into the perhaps eternally inexhaustible abyss of philosophical error, for such a basis proves to be an inadmissible assumption, and merely theological morals in disguise."[53]

Schopenhauer was convinced that Kant had lost touch with the real world when he demands virtually suprahuman virtues in order to live an ethical life. He gave his ethicist colleagues the "advice of first looking around a little at the lives of men," in order to avoid the production of even more ethical theory that resembles Kantian morality.[54]

Indeed, it has been just this that philosophers who are working in the field of applied philosophy have done. They accepted that Kant's concepts still have their merits, but also significant weaknesses (as the preceding paragraphs undoubtedly have demonstrated). One problem, in particular, was that his ethical theory is too general to offer us any guiding moral principles for our daily lives. This has led to more practical theories of autonomy and to strategies that justify interferences with the life concepts and life plans of autonomous individuals on the basis of weak and strong paternalistic concepts.

Before I discuss some of these theories, I will turn to an analysis of John Stuart Mill's theory that would support weak paternalistic interferences. Mill's account of autonomy focuses primarily on the liberty of action. The acts of his autonomous individuals are based on their self-interest. The only reason allowing us to permanently prevent them from doing what they intend to do is that the realization of their interests might harm the interests of other beings. Thus, Mill presents us with a theory that is almost complementary to Kant's theory of the freedom of the will, which is the limits of the individual's freedom to act.

WEAK PATERNALISM: JOHN STUART MILL

John Stuart Mill's (1806-1873) classical essay *On Liberty* discusses problems of autonomy in great depth. However, he never uses the *terminus technicus* "autonomy" but the term "liberty" when he analyzes problems regarding autonomy. His view is basically that autonomy is freedom from coercion by other parties. Autonomy "requires liberty of tastes and pursuits; of framing the plan of our life to suit our own character; of doing as we like."[55] Following Mill, the only limit that allows us to interfere with the intended acts of autonomous individuals is the prevention of harm to others. In Mill's words, "His own good, either physical or moral, is not a sufficient warrant" to intervene.[56]

The consequences of Mill's argument follow the tradition of the Aristotelian view, as the ancient Greek philosopher explicated in his *Nicomachean Ethics*. Here Aristotle argued under the heading "Can a man treat himself unjustly?": "For he suffers voluntarily, but no one is voluntarily treated unjustly. . . . It is not possible to treat oneself unjustly."[57] The German philosopher Arthur Schopenhauer, with reference to Aristotle, comes to a similar conclusion in his fundamental critique of Kant's ethical theory, as laid out in his book *On the Basis of Morality:* "What I do to myself is always only what I will, and consequently it is never a wrong or injustice."[58]

Volenti non fit iniuria (To a person who consents, no injustice is done) is the underlying normative premise of these views. Such diverse philosophers as Mill, Aristotle, and Schopenhauer find a consensus concerning the question of normative judgments about the consequences of such acts. They do not believe that persons cannot hurt or harm themselves, but they claim essentially that, in any case, this is the person's business. No injustice has been done when an individual hurts him- or herself, if this person was beforehand fully aware of at least the possibility of a consequence of this nature and if this person has accepted voluntarily to take the risk.

Aristotle believes that whoever accepts such a risk, accepts the negative consequence and has not been treated unjustly. Clearly, his view is diametrically opposed to paternalistic interventions. This ancient philosopher argues that if you would like to do X, and you do it voluntarily, and you are aware of the possible negative consequences, then there is no injustice done when you hurt yourself because you were aware of these consequences in the first place.

Mill is more interested in another question, that is, under which circumstances is it justifiable to intervene in the individual lives of autonomous beings? His conclusion is that neither the moral or physical welfare of those about whom we are concerned is sufficient to justify coercive measures. According to Mill, it is logically possible that such an individual's plan to commit an act against himself or herself is considered by society to be an immoral act against oneself. However, the consequences would still (by definition) affect only the autonomous individual who made this decision. Thus, Mill differs here from Aristotle and Schopenhauer, both of whom would have argued that there are no immoral acts one could commit against oneself.

Mill is not concerned about whether or not there is any logically coherent way in which immoral acts could be said to be committed against oneself. He could even concede that such acts exist, but he would deny that this constitutes sufficient reason for a coercive interference with the individual life plans of persons who wish to commit such acts.[59] He would argue against all kinds of paternalistic interventions.

As noted earlier, in Mill's view, the only legitimate reason to interfere with individual life plans is to prevent harm to others. In particular, Mill is concerned about the possibility that the moral values of the majority of people in a given nation will be enforced against the convictions of a minority of similarly situated people. His argument against such enforcement is that the evil (or harm) that is caused through such actions—in which the views of the majority regarding the right way of life are enforced against the expressed convictions of a dissenting minority—is greater than the gain.[60] This argument, claims Mill, remains true even when we consider the damage the minority members inflict upon themselves. He does not exclude categorically the possibility that an individual has obligations against him- or herself. What he does, however, exclude, is that society (or even concerned individuals) could have the moral right, or even a moral obligation, to intervene in circumstances in which individuals intend to violate such alleged duties toward themselves.[61] This clearly opposes the Kantian teachings discussed earlier. Mill summarizes his views on this subject as follows:

> What are called duties to ourselves are not socially obligatory, unless circumstances render them at the same time duties to others. The term duty to oneself, when it means anything more than prudence, means self-respect or self-development, and for none of these is any one accountable to his fellow creatures, because for none of them is it for the good of mankind that he be held accountable to them.[62]

Schopenhauer, on the other hand, would stress how important it is to understand that nobody can possibly do injustice to him- or herself. When I decide (on the basis of my individual concept of life) to accept a certain risk to have a chance to receive a certain benefit, then it is obvious that the possible harm (which is part of the deal) is something that was part of what I wanted. Therefore, it cannot be said that I treated myself unjustly or unfairly. Things would be different if somebody else harmed me in a manner to which I had not voluntarily agreed. Nothing else can be meant when Schopenhauer argues that what I do to myself is always only what I will, and therefore it is never an injustice.[63]

Which conditions must be met before a certain entity is, in Mill's view, an autonomously acting being? He excludes categorically the possibility that children or other underage persons could possibly act autonomously.[64] Mill wrote:

> The individual who is presumed to be the best judge of his own interests may be incapable of judging or acting for himself; may be a lunatic, an idiot, an infant: or though not wholly incapable, may be of immature years and judgment. In this case the foundation of the *laisser-faire* principle breaks down entirely. The person most interested is not the best judge of the matter, nor a competent judge at all. Insane persons are everywhere regarded as proper objects of the care of the state. In the case of children and young persons, it is common to say, that though they cannot judge for themselves, they have their parents or their relatives to judge for them.[65]

Furthermore, he excludes persons who are "delirious or in some state of excitement or absorption [that is] incompatible with the full use of the reflecting faculty."[66] Elsewhere, Mill emphasizes that it is "perhaps hardly necessary to say that this doctrine is meant to apply only to human beings in the maturity of their faculties."[67]

For obvious reasons (AIDS is a predominantly sexually transmitted disease), most persons with AIDS are of mature age, and normally, they are not delirious. When we discount those cases of AIDS-related dementia, an important question that is of relevance for the remaining majority of cases remains. Namely, what impact does the knowledge of suffering from a terminal disease have on one's capability to make an autonomous judgment? Undoubtedly, this knowledge will cause some kind of stress in most of these persons. Utilizing this problem, Andrew Shorr has argued that "people facing terminal illnesses appear particularly susceptible to abuse. [They] may reveal a degree of desperation that clouds their judgment and thus, in some cases undermines their ability to make completely autonomous choices."[68]

This is an odd argument because nobody in the real world is able to make completely autonomous choices. Normative premises, religious or political belief systems, and social norms, some of which we internalize unconsciously, always limit our autonomy. Our ability to be completely autonomous in the real world is also handicapped through a variety of other external and internal constraints, such as a lack of knowledge or money.

Given the currently accepted standards, people with AIDS are regarded in all countries as competent, autonomously acting persons. The decision-making capabilities of these terminally ill persons are not in doubt when it comes to management of their day-to-day life. Indeed, they are even considered competent to give informed consent to join clinical AIDS trials. Seemingly, Shorr deems their abilities handicapped only when they decide to make a decision he himself would not make. Judgments of individual competence or incompetence should never be made on the basis of a person belonging to a certain group of people (for instance, sick), but on the basis of an assessment of the particular individual's capabilities.[69]

The respective laws in the United States, Australia, and Germany support this stance. If we accepted Shorr's point of view, I believe we would soon find ourselves on the slippery slope to a situation wherein we could easily claim that people belonging to certain religious groups, or elderly persons, are unable to make fundamental decisions about their lives because they are not fully autonomous.

For the case that is currently under consideration, the question of the participation of persons with AIDS in clinical trials, Andrew Shorr has already decided that the only rational decision can be to join controlled clinical experiments. If a patient decides otherwise, perhaps because he is unwilling to accept the significant risks involved in the participation in such trials, Shorr would conclude that he is perhaps not fully autonomous in his decision-making process. He would argue that this is the case because of the stress caused by knowledge of having a terminal condition.

Let us assume for a moment that persons who suffer lethal illnesses are indeed more likely to test experimental therapies than those who do not, even at a significant risk to themselves. Could this possibly prove that such persons are incompetent? Must the very decision to take a (very) high risk in the face of death necessarily be interpreted as a sign of a panic-related incompetence in the people who are willing to take their chances? This is not necessarily the case. When a person is faced with the inevitable outcome of disease progression, that is death, and this person is given the chance to try an experimental therapy with unknown outcome, then it is obvious that the worst possible outcome would again be the death of this person. All other outcomes would improve the baseline situation of this patient.[70] It seems reasonable to assume that a person with a terminal disease weighs the possible options to act more carefully than a person who is not suffering a terminal disease.[71] This patient is aware that a wrong choice might lead to consequences that are irreversible, given the nature of this particular disease. A person who tries another useless drug against the common cold will not suffer such irreversible consequences.

Buchanan and Brock have proposed to link the necessary standard of competence a patient needs to demonstrate to the expected harms and benefits that would be a likely consequence of his or her intended course of action. They argue that, if a patient wishes to follow a course of action that would lead to a net balance which is substantially worse than that of other alternatives, he or she would have to prove a high/maximal level of decision-making competence.[72] I will discuss their proposal in the Joel Feinberg section of this chapter. For the moment, it shall be sufficient to point out that, even if one accepted such a threshold (which does not rely on a single standard of competence), people with AIDS could still justifiably argue that the net balance of taking an unapproved drug cannot be much worse than what the most likely outcome of taking approved drugs is—death within a relatively short period of time.

In all legal systems of democratic societies, the burden of proof concerning the question of whether a given person is competent or incompetent is on the side of those who claim that this person is incompetent. The hurdles that need to be overcome by those who seek the decision that someone is incompetent are fairly high.[73] The speculations of a bioethicist, such as Andrew Shorr, which pertain to whether being a member of a group of persons with terminal diseases is sufficient to declare persons (when they refuse standard clinical trials) incompetent, will not change this situation, and rightly so. Schorr's strategy seems to be very much ad hoc, since each time a terminally ill patient of a physician such as Shorr chooses to disagree with the physician's advice, he or she will subsequently be declared incompetent.

Mill's position is, unfortunately, not as clear as it might at first seem. On the one hand, he speaks of persons in the "maturity of their faculties." This implies that he considers that the only people who are able to act autonomously are those who are at the height of their intellectual development. On the other hand, in the subsequent two sentences, Mill makes it unmistakably clear that he means to exclude only those persons who have been declared incompetent on the basis of the current legislation.[74] People with AIDS do not belong in this category. They are primarily adults. Some people with AIDS need to be cared for by others, either from a certain stage onward in their illness or intermittently. This requirement, however, is not due to their limited intellectual capabilities, but rather because they develop opportunistic infections. Therefore, they do not belong to the excluded group of persons that Mill had in mind.

Andrew Shorr thinks that human beings with knowledge of their terminal illness are not completely autonomously acting persons. We will see in this chapter's fifth section that we have sufficient reason to reject positions that expect us to respect only fully autonomous decisions. Contrary to Shorr, I find it unlikely, if not impossible, that any human being can ever act fully autonomously. My opinion is in accordance with Beauchamp and Childress, who consider a "substantially autonomous" decision as sufficient to be accepted. Their argument is quite convincing:

> For an action to be autonomous we should only require a substantial degree of understanding and freedom from constraint, not a full understanding or a complete absence of influence. To limit adequate decision making by patients to the ideal of fully or completely autonomous decision making strips these acts of any meaningful place in the practical world, where people's actions are rarely, if ever, fully autonomous. A person's appreciation of information and independence from controlling influences in the health care setting need not exceed, for example, a person's information and independence in making a financial investment, hiring a new employee, buying a new house, or selecting a university. Such consequential decisions are usually substantially autonomous, but far from fully autonomous.[75]

Shorr's argument would lead us to a situation of which, in all probability, nobody would approve. Such a situation would be one in which it was logically conceivable to declare all people with life-threatening or terminal diseases to be incompetent to handle all matters related to their illness after being made aware of their condition. Mill's own condition of being in the maturity of one's faculties is somewhat ambiguous. It is possible to interpret this premise as the required basis of fully autonomous action. It may also be the case, however, that Mill interprets the maturity of a human being's faculties in a legal sense. In this case, to establish that someone is autonomous, it would be sufficient that a given individual is not considered legally incompetent to guide his or her own life. This is a possible account of autonomy that fits into the concept of substantial autonomy, as Beauchamp and Childress have proposed it.

For Mill, the reason for placing such significant value on individual autonomy and self-ruled acts is based on utility. He also claims, however, that "all that makes existence valuable to any one, depends on the enforcement of restraints upon the actions of other people."[76]

He defends his view on individual liberty within the framework of utilitarian reasoning: "I regard utility as the ultimate appeal on all ethical questions; but it must be utility in the largest sense, grounded on the permanent interests of a man as a progressive being."[77]

There is an obvious question in response to Mill's arguments: is it possible to provide a coherent argument that would permit the defense of the principle of utility and still leave nearly complete control over one's own body to the individual, no matter how much he is going to harm himself? A possible interpretation can be found in the principle of the *volenti non fit iniuria*, as has been proposed by Schopenhauer. A person who is aware that the outcome of a certain option to act implies the likelihood of negative consequences is not suffering any setbacks in his attempt to satisfy his or her interests, desires, or preferences.

As Schopenhauer would undoubtedly point out to the complaining gambler who has taken a chance and lost: "What you do to yourself is always what you want, and consequently it is never a wrong or injustice." This argument would allow us to keep Mill's principle of utility as well as his high regard for individual freedom or liberty rights. It seems central to Mill's defense of respect for individual autonomy that he excludes categorically certain types of interventionist measures, among them "paternalistic, moralistic, and gut reactions."[78]

There is also another strategy that can be invoked to defend Mill against the charge of inconsistency which holds that Mill cannot defend such far-reaching individual rights as he proposes, while at the same time insisting on the overall primacy of the principle of utility. To illustrate such a defense, let us imagine a society in which societal or governmental interference in individual lives is permitted/morally justified only when harm to others is imminent. Now, such a society would act in accordance with Mill's idea of the proper scope of individual liberties. The important point is, however, that the maintenance of this society can be defended on utilitarian grounds.

The utilitarian can argue that any transgression of this particular principle of noninterference would, in terms of harmful consequences, lead to an overall negative effect on the social fabric of this society. Following this argument, one can easily admit that it is detrimental to an individual's health to smoke a lot, sunbathe without adequate protection, or drink alcoholic beverages in excess. This alone is not sufficient to make the case for a paternalistic intervention by, for instance, governmental health authorities.

The Millian critic could argue that the negative impact of these interventions, which violate the liberty of autonomous individuals, is much higher than the prevention of certain types of physical harm. Such a society would find itself fairly soon on the road to the totalitarian health state, as Robert Harris has pointed out in a critique of H. L. A. Hart's views:

> One wonders how Hart would limit government's power over the physical well-being of its citizens. The legitimization of physical paternalism could conceivably justify the imposition of a Spartan-like regimen requiring rigorous physical exercise and abstention from smoking, drinking, and hazardous pastimes.[79]

Indeed, the Australian psychologist, Lucy Sullivan, has proposed such measures in the case of beats[80] and bathhouses, in order to prevent AIDS transmissions. She wrote in a letter to the editor of the *Medical Journal of Australia*, "compulsory closing of beats and bathhouses would seem to be a sensible step. . . . In this case, as with speed limits and smoking prohibitions the community has a right to restrict behavior which is costly to both the individual and the society, where it is publicly expressed."[81] I am not so much concerned about the doubtful practicality of this proposal, even though there is reasonable room for skepticism, as Simon Chapman has shown in his response to Lucy Sullivan's letter.[82] It is fairly obvious, though, that the possible positive benefits of Sullivan's proposal, once they have been balanced against the significant intrusion of the health authorities in the lives of autonomous Australian citizens, cannot outweigh the harm done by these draconian measures.

Tom L. Beauchamp hints at another problem with concepts such as this. He asks whether such a strategy would not force us to intervene in order to prevent altruistic people from harming themselves in those situations where the intended act bears a significant risk for the altruist. Beauchamp concludes, "If that is so, then the state is permitted to restrain coercively its morally heroic citizens, not to mention martyrs, if they act—as such people frequently do—in a manner 'harmful' to themselves."[83]

Mill makes two exceptions to his basic principle of intervention only in the case where harm to others is demonstrable. First, it is in his view acceptable (even morally required) to restrict the liberty to act of a normally autonomous individual if this person is unaware of likely and potentially adverse reactions. He uses the following example:

> If either a public officer or any one else saw a person attempting to cross a bridge which had been ascertained to be unsafe, and there were no time to warn him of this danger, they might seize him and turn him back, without any real infringement of his liberty; for liberty consists in doing what one desires, and he does not desire to fall into the river.[84]

Mill would always consider interventions in such situations to be morally justifiable. This kind of interference is considered in today's literature as weak paternalism. As C. L. Ten has pointed out, interference in this case is justified only if there is not time to warn the person about the danger: "It is his ignorance which justifies a temporary restriction of freedom."[85]

If we accepted Mill's interpretation of this type of interference, we could argue that an intervention by the authorities against the declared will of a person who would like to cross such an unsafe bridge becomes immoral if and only if the person in question is aware of the potential consequences of this act. The worse consequence, Mill notes in his scenario, is "to fall into the river."[86] He leaves open whether this translates into a minor inconvenience, such as getting wet clothes, or whether he believes this is also true in the case of a very high bridge where the fall into the water might be lethal.

Prima facie, I think, we have reason to assume that both cases are covered. Mill requests that only one condition be met by the potential bridge crosser, that is, that the individual is informed about the likelihood that he or she might fall. Thus, Mill does not seem to think that the seriousness of the potential harm makes any difference at all in regard to levels of competence required from the person in question.

The second exception to Mill's principle of nonintervention is his claim that nobody should be allowed to give or sell himself or herself into slavery. Such a stance is surprising, since it clearly represents a type of strong paternalism, which would seem irreconcilable with his general opposition to such measures. Mill argues as follows:

> by selling himself for a slave he abdicates his liberty; he forgoes any future use of it beyond a single act. He therefore defeats, in his own case, the very purpose which is the justification of allowing him to dispose of himself. . . . The principle of freedom cannot require that he should be free not to be free. It is not freedom to be allowed to alienate his freedom.[87]

Several attempts by Mill supporters have been made to defend this seemingly incoherent position.[88] Perhaps the most notable of these defenses is that articulated by C. L. Ten in his book, *Mill on Liberty*, where he provides an interpretation of this problematic passage in Mill's work, seeking to defend Mill against the charge of inconsistency. Interestingly, he does not attempt to defend the case for strong paternalism. Rather, he resorts to an interpretation of slavery which concludes that harm to others occurs in such cases and that it is for this reason that it should not be legally recognized.

Slavery, Ten claims, is different from other obligations we accept when we sign contracts in that "the freedom is given up completely and permanently."[89] Slavery is held to be a special situation because we might, at one point, regret having signed the contract that placed us in a state of perpetual slavery. Thus, in accordance with Ten's argument, we might hold that a slave has not agreed to be harmed by the consequences of hard labor, and hence, we can reasonably consider such a case as one in which harm to others occurs. Ten's attempt to defend Mill follows this rationale:

> It is possible that Mill wishes to treat at least some cases in which one person harms another in conformity with an earlier contractual agreement, but against the person's present wishes, as conduct causing harm to others against their will, and therefore as conduct coming within the legitimate scope of legal intervention. If this is the case, then interfering with such contracts is not a paternalistic act but one designed to prevent harm to others.[90]

This argument is unconvincing because an important condition for being able to go voluntarily into slavery (just as crossing the dangerous bridge) is clearly that I have a sufficient awareness of the negative consequences that this decision might imply for me. If I wish at a certain point in time to go into slavery, and I am unaware of the implications of such an act, we would have only another variation of the bridge example in front of us. An intervention under such circumstances would not require a justification for a strong paternalistic act because this act would not meet the conditions of strong paternalism in the first place. However, such an argument was obviously not what Mill had in mind. He excludes the option that free (autonomous) persons should ever be given the opportunity to give up their liberty once and forever, no matter what the consequences might be and regardless of whether or not the acting individual is fully aware of the possible consequences of his or her act. This is also one of the reasons why he does not even attempt to separate cases of malevolent and benevolent slaveholders.

Donald Regan has proposed another defense of the Millian position in his essay, "Justifications for Paternalism." Regan agrees with Mill's central tenet "that the only purpose for which power can rightfully be exercised over any member of a civilized community, against his will, is to prevent harm to others." His central question can be formulated as follows: What is to be said in regard to cases of persons who get hurt and are physically identical with the human being who has given informed consent but who has a new personality and character? He uses the example of a motorbike cyclist who decides to drive without a helmet and subsequently suffers severe head injuries in an accident that could have been avoided if he or she had worn a helmet. "If the cyclist is a different person at a later time from the person he was at the time of his original decision, then the cyclist at the time of the original decision has harmed another person," argues Regan.[91] He admits, though, that even an accident of this kind does not necessarily lead to a different kind of personality for the motorbike driver. He qualifies the necessary changes in the personality of the cyclist in the sense that he is "no longer the sort of person who would ignore his future well-being for the sake of small increments of present utility."[92] Regan limits (in a manner consistent with Mill's example of the passerby who wants to cross the dangerous bridge) the use of coercive means to such cases where a person whose personality changes after incidents such as a motor accident. All those who would insist that it was better to drive without helmet, even after they have had such an accident, should be free to do so.

Regan concedes that a possible (and the only logical) conclusion of his argument could be to question the validity of the concept of individual freedom of action. A central presumption of freedom of action is the possibility to act freely when deciding between a number of choices. This requires, especially if the decision is supposed to be rational, the time to balance the advantages and disadvantages of the available options against each other. If, however, and this is what Regan's argument suggests, we do not have the same personality over time, then we would do better if we were to understand ourselves as a series of personalities, "which correspond to what we used to think of as persons."[93] An important implication of this view is, of course, that we cannot ascribe a certain rationality to one personality, but to a series of personalities. The motive for Regan not to make this final step is, as he claims, of a pragmatic nature: "The reason for my reluctance to take this drastic step is that it would seem to undermine our whole concept of freedom."[94]

My impression is, however, that Regan is unwilling to take his own argument *ad absurdum*. What he really seems to be saying goes as follows: If a person acts at a certain point in time in a certain way, and this act has negative, perhaps irreversible consequences, then it is likely that this person will learn from his or her experiences. It is reasonable to assume that this person, if he or she had the choice a second time, would not try to do the same thing again.[95] Given that every act we do has *per definitionem* irreversible consequences, I wonder what the point of this argument might be—not to act at all anymore? This would not really solve the problem because even the omission to act leads inevitably to irreversible consequences.

As Mill correctly pointed out in another context, "A person may cause evil to others not only by his actions but by his inaction, and in either case he is justly accountable to them for the injury."[96] I think all that remains from Regan's argument is the somewhat trivial realization that we learn from our experiences and that we do better not to repeat the mistakes we have already made. What remains also is the insight that the higher the risk of a certain intended action, the more careful the person should be when balancing the potential risks against the possible benefits. It is undoubtedly the case that others do things we would not do ourselves or that we intend to act in a certain way that others would never even dream of doing. To stop autonomous beings from taking such risks would lead us into a society that polices all our actions.

However, being able to take such risks is a very important part of our attempts to live our own lives. The argument that some (or many) of those who take such risks subsequently regret that they have done so cannot be convincing. After all, they were aware of the risk and decided to take it. The very nature of the concept of risk taking is, of course, that some lose out while others gain. Naturally, there will always be those around who lose out and wish to undo what happened. Those who took their chances and won, however, will be far less likely to complain about the world's injustice.

I consider both Ten's and Regan's arguments unconvincing insofar as they attempt to show the ultimate consistency of Mill's argument against the possibility of making an autonomous choice to go into slavery. It seems to me that Mill is indeed incoherent in this part of his thesis, even though I sympathize with his cause. As we will see later, it is of some importance for my argument to show that this practical conclusion of Mill's does not fit into his views in regard to the individual liberty or rights of autonomous persons to govern their own lives. People with terminal diseases do, of course, run the risk of losing their autonomy irreversibly. Mill's argument against slavery might be misused in this context in order to show that he would not subscribe to the view that such persons should be given the right to risk their autonomy. This risk, after all, seems to be much higher than falling from a bridge and risking saturation by falling into a river.

Quite independently of the claim that Mill is inconsistent at this stage of his discussion of the application of his principle, however, it is important not to forget that people with terminal diseases are *per definitionem* unable to fall back onto a successful standard treatment. It may well be that they can use certain drugs to reduce their suffering or, in the case of AIDS, to prevent certain opportunistic infections, but sooner, rather than later, they will succumb to the illness. This means that the inevitable endpoint of their disease progression is—should they decide not to take the risk of using an experimental drug—the ultimate loss of their autonomy.

Mill's departure from his fundamental principle, which he has so eloquently presented in his introductory notes to *On Liberty*, is inconsistent: "the only purpose for which power can be rightfully exercised over any member of a civilized community, against his will, is to prevent harm to others. His own good, either physical or moral, is not a sufficient warrant."[97] This judgment is supported, for example, by Richard Arneson, who has demonstrated persuasively that Mill's views on the particular subject of slavery are nothing other than an indefensible departure from his otherwise basic stand of weak paternalism.[98]

Proponents of medical paternalism, however, have used his inconsistency in the application of his principle, to claim that "Mill's autonomous agent may not act in ways that would diminish his own autonomy. . . . Autonomy is to be preserved at all cost."[99] Still, a more careful study of *On Liberty* reveals quite clearly, and beyond any reasonable doubt, that the only passage in this work that might support such a view is the one discussed above, in which Mill denies autonomous beings the right to give or sell themselves voluntarily into slavery.[100]

Regan offers another interesting argument that might motivate us to prevent harm to self without violating the general idea of Mill's fundamental principle, which is that such prevention is justifiable only in those cases in which harm to self necessarily implies harm to others. A situation that might warrant this sort of intervention might be the case in which I take the risk of hurting myself in a way that would make me dependent on somebody else's care for the rest of my life. This argument seems reasonable, and I will return to it in the next chapter, which analyzes more closely the conflicts between individual and alleged societal interests.

Another vocal and influential critic of the idea that we should allow autonomous beings to sell themselves into slavery if they decide to do so was Gerald Dworkin. He begins his argument by rejecting the following view of his own, which he proposed in 1968: "The main consideration for not allowing such a contract is the need to preserve liberty of the person to make future choices."[101] His reason for abandoning this argument in 1988 is persuasive:

> [This view] raises the following theoretical problem. There is nothing in the idea of autonomy that precludes a person from saying, "I want to be the kind of person who acts at the command of others. . . . My autonomy consists in being a slave." If this is coherent, and I think it is, one cannot argue against such slavery on the grounds of autonomy.[102]

However, Dworkin, like most others, feels bad about the idea that some people might really sell themselves into some slaverylike situation. Thus, he offers a different strategy to shift around the problems facing his 1968 position. Dworkin proposes that we take a closer look at the social consequences the permission of slavery might have for all of us. Would we want to be the person who is sending a slave who ran away from his or her owner back to the owner, because the society in which we live respects that kind of contract?[103]

Dworkin's concern is very different from Mill's. He argues that we should not have the opportunity to sell ourselves into slavery, not because of the harm one does to oneself, but because of the harm done to others, in particular, the harm done to the social fabric of a given society. I find this argument convincing because it is true that, one way or another, people who lived in such a society would have to confront the problem of slaves who run away from their owners or slaves claiming that they have been variously abused by their respective owners, etc.

People who find the idea of slavery appalling and demeaning to human dignity will be especially distressed in such situations. There is, however, no analogy to be drawn to the situations of people with AIDS which would support the conclusion that they cannot be allowed to risk their autonomy by using experimental drugs in a last-ditch attempt to save their individual lives. On the contrary, we might well ask ourselves whether we would want to live in a society that has legislation in place which forces us to hand over to the police terminally ill people who plan to buy and use experimental drugs or who have actually done so.

It seems that Mill's position might best be characterized the following way: An adult, who is not intellectually handicapped, is entitled to perform any act, as long as he or she is aware of the consequences of this act and as long as these acts do not affect other parties significantly or in an overwhelmingly adverse manner. If the knowledge of such an individual is insufficient, a second party may justifiably interfere with the person's intended acts temporarily, in order to render the individual informed about the likely harmful effects of these acts.

If the autonomous being decides subsequently that he or she wishes to proceed with the plans, we have no ethical entitlement to intervene again, even in such cases where we consider the intended acts to be imprudent. If we were to assume that the intention to buy and use unapproved drugs, in cases of persons with terminal diseases, had no significant negative effects for anybody other than the patient who wishes to take the risks involved,[104] then Mill's reply to the more fundamental of the two problems I have set out in the introduction should follow these lines: the state has no moral right (or even obligation) to prevent persons with terminal illnesses from buying and using unapproved drugs, as long as they are aware of the potential adverse consequences that taking this risk implies.

In order to prevent persons with lethal diseases from taking unknown risks, it is necessary to keep them informed about the magnitude of these risks.[105] An obvious argument against this claim is that its acceptance would render it impossible to give consent to a risk of unknown magnitude. Robert Goodin, for instance, argues that "consent counts—morally, as well as legally—only if it is truly informed consent, only if people know what it is to which they are consenting."[106] Among other conditions, he requires that we "can state the probabilities."[106] This is *per definitionem* impossible in the case of experimental drugs that have not been scrutinized in either preclinical or clinical trials.

The bridge example presented by Mill affords a different interpretation. It seems to support the view that autonomous beings can actually accept risks of an unknown magnitude as long as they are aware of the fact that such risks exist. Mill argued that we are only temporarily justified in preventing people from crossing the bridge. Once they are informed of the risks they are facing, we have no moral justifications for preventing them from crossing the bridge, if they still wish to proceed with their original intention.

Given that, in this example, we are faced with a risk of unknown magnitude (after all, a whole range of outcomes are imaginable, from mere saturation to being injured, drowning, being eaten by crocodiles, etc.), it seems quite obvious to me that Mill thinks it is reasonable to allow autonomous individuals to accept risks of an unknown magnitude as long as they are aware of that situation. The only condition he required is that the paternalist has given them a rational assessment of the (ultimately unknown) risk they are taking when they go ahead with their original plans.

Surprisingly, Mill has taken a clear stand in *On Liberty* in regard to one of the questions I have posed in the introduction. This problem remains at the center of a significant dispute among AIDS activists on the one hand and the clinical research establishment on the other. He wrote: "No one but the person himself can judge of the sufficiency of the motive which may prompt him to incur the risk . . . he ought, I conceive, to be only warned of the danger; not forcibly prevented from exposing himself to it." It is, therefore, only consistent that Mill also argued it is illegitimate "to require in all cases the certificate of a medical practitioner."[107] As George W. Rainbolt proposed, "Mill might argue that if we (1) require insurance, (2) put on warning labels, and (3) train our pharmacists to spot people who were unable to make sufficiently voluntary decisions, then prescription laws would no longer be possible."[108] Seemingly, today's AIDS activists have a strong, influential ally in John Stuart Mill and his disciples.[109]

WEAK PATERNALISM: JOEL FEINBERG

Joel Feinberg (1926-19__) leaves his readership in no doubt that he considers himself a liberal in the Millian tradition.[110] Liberalism, as he interprets it, is the view that only harm and offense principles "state good and relevant reasons for state coercion by means of the criminal law."[111] His philosophical magnum opus consists of a four-volume inquiry, *Moral Limits of the Criminal Law*. Two of these volumes are related to the question of whether at all, and if so, under which circumstances, state interventions are legitimate to prevent harm to self and harm to others.

His analysis of several autonomy-related concepts begins quite appropriately with a brief description of how a paternalistic intervention is perceived by those into whose lives the paternalist intervenes:

> Their grievance is not simply that they have been unnecessarily inconvenienced or "irked," but rather that in some way they have been violated, invaded, belittled. They have experienced something analogous to the invasion of their property, or the violation of their privacy.[112]

The accuracy of Feinberg's interpretation is clearly emphasized when we consider the following example of the feelings of an AIDS patient who wishes to take unapproved drugs. Martin Delaney, a director of the U.S.-American AIDS activist group Project Inform (San Francisco), quoted this patient in his address to the twenty-sixth annual meeting of the U.S. Infectious Diseases Society:

> It is as if I am in a disabled airplane, speeding downward out of control. I see a parachute hanging on the cabin wall, one small moment of hope. I try to strap it on, when a government employee reaches out and tears it off my back, admonishing, "You cannot use that! It does not have a Federal Aviation Administration sticker on it. We do not know if it will work."[113]

Feinberg believes there are at least four closely related meanings of autonomy: "It can refer either to the *capacity* to govern oneself, ... or to the *actual condition* of self-government, ... or to an *ideal of character* derived from that conception, or to the *sovereign authority* to govern oneself."[114] The capacity to govern one's own journey through one's life depends clearly upon a necessary condition, that is, the capacity to govern oneself. Everybody above a certain threshold of competence has, in most countries, the de jure right (capacity) to govern his or her own life. Feinberg is quite right when he argues that certain persons above this threshold may use their autonomy for foolish or reckless actions but that they are certainly entitled to these actions as much as, for instance, a nation may be governed stupidly.

Feinberg clearly analyzes the personal sovereignty of the autonomous individual as analogous to that of the sovereign state. Indeed, he believes that "a genuinely incompetent being, below the threshold, is incapable of making . . . choices."[115] Feinberg interprets autonomy as an actual condition. It entails a number of virtues that derive from a concept of self-determination. Among them, evidently, are self-possession, -identity, -legislation, -fidelity, and -reliance. For Feinberg, as for the subsequently analyzed work by Gerald Dworkin and Robert Young, the condition of authenticity seems to be crucial for autonomy.

In this regard, his argument sounds familiar: an autonomous person is not the mouthpiece of others but is representing his or her own (authentic) values. Feinberg also realizes that even the authentic person cannot possibly be entirely self-made because "if we take authenticity to require that *all* principles are together to be examined afresh in the light of reason on each occasion for decision, then nothing resembling rational reflection can ever get started."[116] Thus, he correctly notes that there can be no ex nihilo creation of the habit of rational reflection—a point that has also been made by Dworkin and Young. The argument is made, however, that the ability to reflect upon those values and ideas we have been taught or otherwise given while we were raised (for instance, by our parents or in school) and the ability to change them at will or to retain them renders those reflected-upon values and ideas our own. In my analyses of Young's and Dworkin's account of individual autonomy, I will argue that the criterion of authenticity is impractical and, hence, often not applicable.

Joel Feinberg presents us with a helpful example of an inauthentic person, which is useful insofar as it provides some evidence for my view that he would have to consider all those people with AIDS who do not join the rush into the latest fashionable drug in AIDS therapy as authentic persons. Feinberg writes:

> The inauthentic person of this type is essentially the manipulated consumer. He has no taste in music or clothes except for what is fashionable this season. If blue flatters his complexion while green makes him sallow and sickly, yet green is "in," he will buy all green shirts, aesthetic considerations be damned.[117]

Now, clearly, it is not entirely impossible that the decision to wear those clothes, which are currently "in," is actually an authentic decision. It would only be necessary that the consumer makes a conscious decision to buy and wear those clothes that are currently fashionable in order to please his friends, family, or whatever. In this case, he would clearly not be the manipulated consumer Feinberg has constructed, but a person whose authentic values have motivated him to balance harms and goods of a strategy to adhere to certain fashionable clothing rules, against wearing what he really likes (for instance, clothes that are considered by all of his close friends to be revolting). Thus the decision to wear what is currently "in" is not necessarily inauthentic. The decision's degree of authenticity depends upon the reasoning behind the choice of clothes. The same is true for a person with AIDS who chooses not to apply what is currently "in," for instance protease inhibitors plus antiretrovirals, but take his chances with certain alternative treatments, which might require him to live a healthy life and to not touch any pharmaceuticals at all.

Of course, one could argue that it is logically possible that a person with AIDS has decided to use "alternative therapies" (defined as all those therapies not approved by established AIDS research as useful in the treatment of opportunistic infections or the underlying immunodeficiency) and is nevertheless the inauthentic, manipulated consumer Feinberg has in mind.

A supporting example for this claim might be where the decision to join networks of alternatively minded persons with AIDS was not conscious but was perhaps due to the influence of a lover or fellow person with AIDS. I will not attempt to deny that such manipulations may take place. I would think, however, that unless empirical evidence to the contrary is provided, we can assume that these cases constitute a very remote minority. I base this claim on the logic that it would seem unlikely that many people who face essentially irreversible decisions about the course of treatment for a life-threatening disease, will follow the advice, even of a beloved person, unless they are convinced that there is some advantage inherent to that proposal. After all, it is unlikely that they receive only the advice of that particular "manipulator," given the highly probable offering of advice from primary care physicians and perhaps even principal investigators and AIDS organizations. Many of these official advisors will go to great lengths to make the case for established standard procedures.

If one looks in the gay papers in any country of the Western world, newsletters, brochures of AIDS organizations, information from health ministers, etc., almost everybody makes the case against alternative experimental therapies. This current situation is vastly different from the beginning of the epidemic, when no reliable treatments for many opportunistic infections were available. In addition, the few newsletters of alternative healing groups that do exist indicate that it would be quite difficult to become manipulated by such papers in the first place.

This may be for the simple reason that these papers are unappealing, insofar as they have a layperson's format and design. Such production is an obvious disadvantage when compared to the glossy materials produced by certain pharmaceutical companies, which try to convince persons with HIV or AIDS to join certain trials in which they have vested interests. Also, we should not forget the millions of dollars for AIDS education that are readily available for those publications and campaigns which support standard procedures. I have provided evidence elsewhere that materials by pharmaceutical companies and AIDS organizations are manipulative.[118]

Given that almost all of the information coming out of the established governmental and academic research institutions would lead patients to join their trials, it would seem that a decision against such a move is more often than not an authentic move. (The prudence of such a decision may be questionable, but that shall be of no concern to us at the moment.) Therefore, following Feinberg's account of authenticity, it seems clear that the decisions people consciously make against the predominant views of the right course in AIDS therapy are at least authentic. In fact, one might more appropriately ask whether those patients who follow uncritically the proposals put forward by our established research institutions should, at least prima facie, be considered as nonauthentic persons, given that they are more or less "habitual and uncritical conformist[s]."[119]

Following the Kantian notions of self-legislation and hypothetical rational consent discussed earlier, Feinberg argues that it is imaginable that a person has no moral convictions whatsoever and is basing his or her actions entirely upon otherwise motivated policies. In such a case, this person's dispositional autonomy is only handicapped through the lack of moral authenticity of his or her actions. Feinberg claims that those persons who "most conspicuously fall short of *de facto* autonomy are . . . those whose 'morality' is a mindless reflex."[120] (Given that this is an important empirical claim, it is surprising that Feinberg offers no references to support his view.) The autonomous person's life is shaped by his or her own moral convictions. "To summarize: the morally authentic person *has* moral principles, and they are *his own* principles, but that does not imply that his will is their source or ground, or necessary for their objective validation."[121]

Feinberg defines (weak and strong) paternalism in his article "Legal Paternalism," suggesting that legal paternalism justifies state coercion to prevent people from harming themselves or to lead them with or without their consent toward their own good.[122] In his assessment of the justifiability of paternalism, he rejects the extreme views of either side. Thus, he agrees with neither the claim that the protection of a person from himself or herself is always an ethically acceptable motive for interference, nor that it is never valid grounds.[123]

In interpreting and defending Mill, Richard Arneson has argued that people have, "in the sphere of self-regarding action . . . the right to make their own mistakes."[124] Although Feinberg seems to accept this proposition, he takes exception to Arneson's analysis of our responsibility for the mistakes we make. He correctly realizes that both Mill and Arneson face the difficulty of distinguishing between mistakes that reduce or diminish voluntariness (which is a necessary precondition for an autonomous action) and those that do not.

For my discussion, however, this dispute is barely relevant because people with AIDS usually do not have mistaken views about the status of the experimental drugs in which they have put their hopes. They are well aware of their limited knowledge about the impact the drug will have on their health. If they (mistakenly) believe that the drug will save their lives, the situation would, of course, change. Still, I am not aware of a single case in which a person with AIDS was not aware of the risks involved in taking experimental, unapproved drugs. However, I would agree that, in cases of mistaken factual judgments made by such patients about the true nature of the risk they take, their intended action is rendered nonautonomous, just as with the person who wishes to cross the bridge in Mill's example.

Perhaps a hypothetical example will help to illustrate this claim. Imagine that a person with a terminal disease wishes to take an unapproved experimental remedy, and he claims (in the absence of any supporting empirical evidence) that it will cure him. In such a case, I think we have a valid, weak paternalistic reason for intervention. However, even in such situations, we have to be very careful, since it is often difficult to assess what constitutes supporting empirical evidence. In order to demonstrate the complexity of such a judgment, I will refer to the real-life example of what happened with a widely used underground drug, peptide T.

A number of people with AIDS claim that peptide T reduces the painful effects of peripheral neuropathy. A trial was undertaken, however, that concluded that this experimental agent is no better than a placebo in the treatment of this particular condition. Hence the comment, "The drug's top-selling status has been fueled by widely repeated anecdotal accounts of relief from AIDS dementia and cognitive impairments, of increases in the quality of life variables, and indeed, of relief from the pain of peripheral neuropathy."[125] The relevant study in this example showed 35 percent improvement after three months in both placebo and drug arms, indicating that the observations of improvement were not caused by peptide T.

The validity of this study was questioned, however, when proponents of peptide T claimed that blood tests demonstrated that many persons in the placebo group had also used peptide T. Unfortunately, this claim could not be independently confirmed. Perhaps understandably, due to the reported anecdotal cases where peptide T was supposedly helpful and because of the fact that it has virtually no side effects, peptide T remains a best-seller in the AIDS underground. Representatives of New York City–based nonprofit buyer's clubs, which sell this drug, claim that they have "anecdotal reports of people with lower pain levels who have success in using peptide T to control it." Moreover, peptide T has been shown to reduce the levels of the cytokine TNF-alpha, which is associated with AIDS disease progression. As a matter of fact, "despite the negative study results, seventeen study participants have elected to continue taking the drug. . . . Several hundred persons with HIV infection and AIDS continue to purchase peptide T on the underground."

Thus this example highlights an important problem; it is not always clear what the real effects of a certain agent are, even though standard drug trials have been undertaken. Hence, it is only reasonable that people with AIDS may justifiably have doubts about whether or not claims about a certain drug are correct. My specific concern in raising this example was to illustrate the difficulty in charging that persons with supposedly "mistaken" conceptions about a drug's properties (as indicated by their nonacceptance of a certain trial's results) are rendered nonautonomous. This difficulty arises because the empirical data are rarely conclusive and open to a number of interpretations, thus confounding the issue of which interpretation might be "mistaken."

In the context of peptide T, I would maintain that the relevant people with AIDS are not nonautonomous just because they doubt the validity of a certain drug trial. Those wishing to establish this case need to provide more evidence for their alleged incompetence than a mere reference to a trial published in a refereed journal. An unbiased article (in the New York City Gay Men's Health Crisis magazine *Treatment Issues*) about the currently known pros and cons of this experimental agent concludes with a warning: "Those who are currently considering the use of peptide T are advised: caveat emptor (let the buyer beware)."

The type of situation with which I am specifically concerned in this book, however, is different from that just described. The former is one where patients with a terminal disease wish to take a certain unapproved drug because they believe they have some indications (which are to them sufficient) that the drug might help. I wish to emphasize that this is a much weaker claim than one which would hold that the drug will definitely cure the patient. Moreover, the weaker claim implies that the patient has realized the uncertainty surrounding the outcome of this experiment. It indicates that he or she has a substantial understanding of the risks involved in this decision. Such an understanding might render his or her decision one that can justifiably be called autonomous because he or she has weighed the potential risks and benefits of the alternative medication "according to [his or her] own values and [has chosen] the alternative that best promotes those values."[126]

George J. Annas has repeatedly misrepresented this situation of people with AIDS who have decided to take the risk of using an experimental agent. He claims people with AIDS, AIDS activist groups, and even the U.S. FDA "seem so intent on denying death that they believe it can be magically prevented with unproven drugs."[127] Contrary to Annas, I would argue that the obvious conclusion to be drawn, when one takes into account the aforementioned quotes and newsletters from actual people with AIDS, is that they are well aware that their imminent death won't be magically prevented by unproven drugs. What they have also realized, however, is that doing nothing while waiting for a drug to be developed at some point in the future will almost surely entail inevitable death. Hence, both these patients and AIDS activists have come to understand that they either take a risk—one they would not take if the odds were not so clearly stacked against their survival—or die. They are well aware of their limited knowledge and of the fact that the experimental drug is unproven.

A survey of the relevant scientific literature reveals that a situation in which we do not have certain knowledge about the impact of a disputed (experimental) drug is indeed the more common one. Feinberg argues that, in such a case, where the facts are scientifically controversial, the person who wishes to apply the experimental drug should be entitled to do so, as long as he or she is aware of running a risk of unknown magnitude (due to the controversial nature of the drug, reflected in the scientific dispute about its impact on humans who choose to take it). He concludes his discussion of a couple of examples thus: "we seem to have a case for special dispensation from an equity board. Dr. Doe, it would seem, does have a right, in the absence of sure knowledge, to 'make his own mistake' about facts that are scientifically controversial."[128] The fact that Feinberg has chosen to make the person who wishes to take the chances a chemist, who has undertaken experiments on the drug himself, does not make an ethically relevant difference. He is a professional (described as a maverick) and wishes to take a certain drug, which he believes will have a positive effect. The majority of his colleagues deny this emphatically. Therefore, our professional (no matter how strong his indications may be) could be mistaken in his judgment of the risks involved. The chances are actually quite high that he is wrong, given that the majority of his peers and colleagues believe him to be wrong. But again, he knows about their doubts and counterarguments and insists that the risk is worth taking.

Clearly, his situation is, in epistemological terms, not at all different from the situation of a patient suffering from AIDS who wishes to take a certain experimental drug that he or she believes can, for example, prevent a usually fatal opportunistic infection of the lungs. Even though our patient is not a professional scientist (such as Feinberg's Dr. Doe), he or she still does not know what the outcome of the experiment will be. Just like Dr. Doe, he or she has more or less valid convictions about the drug. Some of them may even be as approximately scientifically controversial as those voiced by Feinberg's scientist.

Elsewhere in *Harm to Self*, Feinberg argues that the state has the right to intervene paternalistically "if a layman disagrees with a physician on a question of medical fact," because "the layman [can] safely be presumed wrong [and] the state intervenes to protect him not from his own free and informed choices, but from his factual ignorance."[129] This may be correct in the case of some approved drugs about which a given patient may have wrong factual ideas. However, this argument does not apply to experimental drugs because the medical facts are by definition unclear. I have indicated before that it is crucial that the patient has the relevant substantial metaknowledge about the uncertainties involved. As C. L. Ten correctly noted:

> Another type of voluntary risk taking, for which metaknowledge alone is sufficient, is present in cases where drug users, knowing that they are ignorant about drugs, also know that the relevant first-level knowledge is currently unavailable to anyone. If they are still prepared to use drugs under these circumstances . . . , then their decision can be regarded as sufficiently voluntary. Others may not share their attitude towards risk, but this only reflects differences in the evaluation of risks.[130]

Feinberg argues that a reason for the state to intervene is that "the risk [somebody wishes to take] is extreme, and, in respect to its objectively assessable components, manifestly unreasonable to the point of suggesting impaired rationality."[131] Feinberg differentiates between irrational choices of incompetent agents and unreasonable choices of competent, autonomous individuals. He concedes that "part of the point of our calling such choices 'unreasonable' is to suggest that they reflect judgments of comparative worthwhileness that we would not make were we in the chooser's position."[132] The "we" in Feinberg's interpretation implies that we share the potential actors' authentic values. Feinberg chooses quite uncontroversial examples to underline his case, such as the unreasonableness of cutting off one's arm with a power saw, which risks bleeding to death, in order to cure an infected finger.

A similar view is put forward by Buchanan and Brock, who have argued that "the standard of competence ought to vary in part with the expected harms or benefits to the patient in accordance with the patient's choice."[133] In essence, they argue, as Feinberg does, that choosing a high risk is indicative of low competence (or maybe even incompetence) or suggestive of impaired rationality.

Mark R. Wicclair countered this argument quite persuasively by pointing out that "there is a danger that standards of understanding, reasoning, and so forth will be set arbitrarily and unattainably high by those who believe that paternalism is justified when perceived risks are great."[134] Indeed, if we consider our AIDS example again, it is clear that the risk-related threshold for competence becomes quite a dangerous instrument in the hand of de facto strong paternalists. The risk of harm to oneself involved in the taking of an unapproved drug may indeed be tremendous (or at least of an unknown magnitude). Feinberg, as well as Buchanan and Brock, refer to some sort of "objective standards" that may be violated by the risk-taking person's choice. In the words of Buchanan and Brock:

> The presumed net balance of expected benefits and risks of patient choice in comparison with other alternatives refers to the physician's assessment of the expected effects in achieving the goals of prolonging life, preventing injury and disability,

and relieving suffering from a particular treatment option as against its risks of harm.[135]

Statistics may even exist to prove that, in most cases, the patients' chances for a positive outcome are remote and that the majority of them will be worse off than they were before taking the experimental drug. Imagine, furthermore, that the authentic values of our person with AIDS led him or her to a pre-AIDS life, which is considered by many as one where all possible risks to life and health have been avoided. Now this person is facing a life-threatening disease. Perhaps, in his or her pre-AIDS life, he or she would have never even contemplated the taking of experimental drugs in order to prevent any of the potential harms that might be associated with taking such drugs.

Let us assume that the individual was aware of the great risks involved in taking such agents. Circumstances, however, have greatly changed. In a poster on science.medicine.aids, an international electronic "moderated" (the Internet version of peer review) newsgroup on the Internet, a person with AIDS made exactly this point to the scientists subscribing to this Internet newsgroup: "However irritating anecdotal medicine may be to trained scientists, I feel we just do not have the luxury to ignore it and dismiss it all out of hand as 'moronic' or 'quackery.' We are running out of time to wait for basic science to provide all the answers."[136]

Dan Brock seems to support this point of view, when he argues that "self-determination is not only important when patients are choosing wisely; it also protects at least some bad or defective choices."[137] Undoubtedly, the decision to take unapproved drugs can be a bad decision, but it can intelligibly be argued that it is a bad decision among a variety of other less than ideal (or bad) choices.

Something interesting has happened within AIDS activist groups, which were, until very recently, united in their fight against the federal AIDS bureaucracy in the United States. Some activists have come to realize that the trials, which have taken place thus far, have presented very fragile evidence for or against certain drugs, such as AZT. The largest and longest running of AZT trials, the Concorde trial, failed to find any survival benefit from the use of this drug in the early stages of disease progression. This result has led some activists to question whether or not they should accept that old-fashioned kinds of placebo-controlled clinical trials, which run over long periods of time and include many patients, are the only solution to this problem.

In New York City, the relatively new activist group TAG (Treatment Action Group) has come to conclude that LSTs (large simple trials) are the only possible answer to the uncertainty caused by short-term trials, some of which have even been terminated prematurely after some clinical surrogate markers seemed to indicate that the experimental drug was better than the placebo. It has been pointed out in the literature, however, that those HIV-positive persons who propose such a strategy are mostly in the early stages of disease progression and have a realistic chance of benefiting from the results of such trials. Many AIDS activists have been highly critical of the TAG proposal. Brie Salzman, for instance, criticized this strategy:

> "A Large Simple Trial of antivirals will cramp the pipeline," he said. "Just getting it off the ground will suck up efforts that should be looking for newer, better drugs. . . . *Too many of us just do not have time* [to wait] *for hard data.*"[138]

The Swedish philosopher, Torbjörn Tännsjö, would perhaps disagree with Brie Salzman on this point. He wants paternalism to be banned only when patients succeed in convincing their physicians on one or both of the following counts: they must, in his view, be capable not only of exercising a judgment of their own, but must also be able to reach "a wise judgment" (which is actually determined by what the physicians consider wise).[139] In articulating this argument, Tännsjö is obviously taking a stand that has practical conclusions similar to those that I have criticized in this section. In order to make a competent judgment, he requires the physician to meet an unrealistic standard: "We had to face the question of whether the trial was in the best interest of the patients. We had to put ourselves in their shoes, in order to make a rational assessment of what the trial would mean to them."[140]

The difficulty with this requirement, however, is the assumption that such a substitution is possible. I would argue that it is not possible that a perfectly healthy person (a physician treating this patient, for instance, or a moral philosopher such as Tännsjö) or even a relatively healthy person with AIDS who is in the early stages of the disease, could possibly imagine what it means to die the death that a person in the end stages of this illness is going to face. We may be able to judge the reasonableness of such a person's intended actions when we are in a similar situation (if the person is healthy, and we are healthy, or if the person has AIDS and we have AIDS, with the occurrence of comparable opportunistic infections). Our judgment must be questioned, however, when we are comparatively healthy and the people over whose intended action we have to judge are dying. What might look reasonable to such people could well seem entirely unreasonable to us.

I contend that the risk magnitude that Feinberg, Buchanan, and Brock mention is irrelevant. Thus, I maintain that there is not a proportionate relationship between the height of the risk and the height of the threshold of "reasonableness" required by the patient. According to my argument, however, it is crucial that the patient demonstrates (if the risk is significant) one way or another that he or she understands the risks involved and is not making any demonstrably wrong factual claims.

To avoid biases toward patients behaving in a way that we find reasonable (joining standard clinical trials), we should ensure that we test patient standards of knowledge in cases where we do agree with the choices being made, as well as those in which patients are making choices contrary to those we would endorse. It appears that Buchanan and Brock would take issue with this argument, since they claim that "just because a patient is competent to consent to a treatment, it does not follow that the patient is competent to refuse it, and vice versa."[141] This is a direct result of their previously criticized idea that the standards of competence required should be risk related. Such an argument would inevitably lead to a situation where patients who adhere to standard procedures and who join clinical trials as advised by physicians would need to demonstrate a much lower level of competence than those who choose not to do so.

Compliance with standard procedures is clearly rewarded with lower standards necessary to prove competence. This situation is unjust. If one is a less compliant patient, and if one rejects the standard treatment proposals, the medical establishment will declare the risk of taking another direction by using experimental drugs tremendous. It should be noted here that the standard treatment proposals that are currently available lead inevitably to the death of people with AIDS.

Consequently, following Buchanan, Brock, and also Feinberg, the level of competence has to be high (in fact, the required standard might be impossible to reach). Recent research, however, has demonstrated how important such a competence test is, especially for those who wish to comply with standard procedures.[142] The simple signature of a given competent patient on an informed consent form is not sufficient. Many of those patients have a therapeutic misconception, which incorporates the "denial that there may be major disadvantages to participating in clinical trials that stem from the nature of the research process itself."[143] Indeed, as discussed next, even those who do not succumb to this particular misconception have been shown to have significant problems in their understanding of what is really happening.

Michael Kirby provides the following example:

> Within one day of signing consent forms for chemotherapy, radiation therapy or surgery, 200 cancer patients completed a test of their recall of the material in the consent explanation and filled out a questionnaire regarding their opinions of its purpose, content and implications. Only 66% understood the purpose and nature of the procedure, and only 55% correctly listed even one major risk or complication. . . . Only 40% of the patients had read the form "carefully."[144]

Michael Kirby's paper and that published by Paul S. Appelbaum and associates are just two examples of a number of studies which suggest that some sort of test is necessary in order to establish whether the patients have a substantial understanding of the technicalities involved in joining a double-blind, placebo-controlled, randomized clinical trial, or any other sort of clinical trial that involves a significant risk for their future well-being.[145] Wicclair agrees with this suggestion for another reason, which is related to his critique of another problematic aspect of risk-related thresholds of competence. He writes, "There is the danger that decision-making standards will be set so low when patients concur with the recommendations of health care professionals that they will be classified as decisionally capable, regardless of their mental status."[146]

So, where does this leave the final requirement mentioned by Feinberg, namely, the responsibility for one's actions? Feinberg agrees with Arneson that, if an autonomous person makes a voluntary choice, he or she has to take on the "responsibility for all the foreseeable consequences to himself that flow from this voluntary choice."[147] It remains to be seen whether this view of responsibility would lead Feinberg to argue that a person who took experimental drugs and harmed himself or herself, to the degree that he or she needs assistance from others in order to recover, is not morally entitled to that help. I shall discuss this question in Chapter 3. In contrast, nonautonomously acting people (for instance, those who are incompetent) cannot possibly be held responsible for their deeds. Feinberg's ideal of an autonomous person is that of an authentic individual whose self-determination is as complete as it is consistent with the requirements that are implicated in being a member of a society.

WEAK PATERNALISM: GERALD DWORKIN

Gerald Dworkin's theory of autonomy can be found in a collection of essays on related subjects, published under the title *The Theory and Practice of Autonomy* in the Cambridge Studies in Philosophy series.[148] Dworkin presents us with a rather weak, contentless, and abstract concept of autonomy, which he justifies in the following quote: "[P]eople can give meaning to their lives in all kinds of ways. . . . Any feature that is going to be fundamental in moral thinking must be a feature that persons share. But any substantive conception of autonomy is not likely to be shared."[149] He defines autonomy as a "second-order capacity of persons to reflect critically upon their first-order preferences, desires, wishes and so forth and the capacity to accept or attempt to change these in light of higher-order preferences or values."[150]

The main thrust of this characterization is identical with that proposed by Harry G. Frankfurt in 1971.[151] According to this conception, human beings who possess these kinds of capabilities and begin to use them in their daily lives start to "live their own lives" in the real meaning of this phrase. They begin to define their own nature and try to give coherence and meaning to their lives. In short, they "take responsibility for the kind of person they are."[152] The disposition to reflect upon themselves allows them to come to solutions and decisions that are truly their own.

These decisions are authentic decisions. They can (try to) live their own lives in accordance with their own values. As Ruth R. Faden and Tom L. Beauchamp pointed out in a critical response to Dworkin, "Authenticity is the most important condition in Gerald Dworkin's influential theory of autonomy."[153] This is the case because "it makes no difference how one comes to have desires . . . but only whether on reflection one accepts them, and thus makes them authentic." They persuasively point out that "people frequently act in ways that represent departures from their 'stable' values. Some of these acts are autonomous. The values of such persons can be in transition, or their values can be poorly formed or irrelevant to a particular situation."[154]

The level of reflection attained will undoubtedly differ among people, depending on a multitude of factors, such as their psychological condition or their education. Dworkin is certainly justified in his concern that such strict criteria for autonomy might be developed that only philosophy professors would be considered autonomous. Thus, in his view, a farmer is autonomous when he is aware of why he is living the life he lives and is capable of changing the direction of his life after he has reflected on it, or of consciously deciding to retain the direction of his life.

Respect for individual autonomy is of special importance in the health care context. As Dworkin puts it, "Because my body is me, failure to respect my wishes concerning my body is a particularly insulting denial of autonomy."[155] His aforementioned discussion of the Millian position on slavery suggests that Dworkin would reject strong paternalistic interferences. He admits that the only plausible reason for an interference in the lives of people who wish (voluntarily) to become slaves is the potential harm done to the quality of living of the rest of the society, which finds this wish appalling. In other words, Dworkin is anxious to refute the suspicion that he might favor strong paternalism under certain circumstances.[156]

On the other hand, he also emphasizes that paternalism which helps to benefit the individual body (or to prevent harm to it) could be justified. He defines paternalism as the "interference with a person's liberty of action justified by reasons referring exclusively to the welfare, good, happiness, needs, interests or values of the person being coerced."[157] His attempt to ethically justify (weak) paternalistic interventions follows from the fact that he places autonomy in a hierarchical order where it is only one value among others. Hence, autonomy is not given the power to override all other values that would be necessary to establish it as the absolute value.[158] It could well be, Dworkin believes, that we have to sacrifice autonomy in order to further other values, such as individual dignity, health, well-being, integrity, and security. Dworkin argues for four areas in which paternalistic interventions seem to be ethically justifiable. These areas comprise: (1) emergency, (2) incompetence, (3) waiver, and (4) therapeutic privilege.

Dworkin's first three examples seem to be relatively unproblematic:

Imagine a life-threatening emergency situation. We are unable to obtain informed consent from the patient to perform a medical intervention in order to save his or her life. Of course, it would be in the best interest of this patient, who might be unconscious or in a state of shock, to save his or her life. This is due to the somewhat trivial fact that most people enjoy being alive and would prefer, generally, to live a life they consider worth living, rather than not live at all. Therefore, in emergency situations where physicians are unable to obtain informed consent from a patient, they have sufficient reason to believe that this person wishes to be saved even if he or she has not explicitly said so. In the case that the person is unhappy about this interference, he or she has plenty of time, post-intervention, to end his or her life in a way he or she deems appropriate.

Such a situation provides a classical case of justified weak paternalistic intervention. The function of performing the medical procedure is clearly to prevent the permanent loss of our autonomy "by authorizing health care when we are not able to do so."[159] The assumption Dworkin is making is, of course, that the person in question will (in most cases) agree to the action taken by the paternalist, after his or her decision-making capacities have been restored.[160] Although his position seems very reasonable, there are limitations to its relevance. As Heta Häyry has warned, the situation Dworkin has described is an exceptional one. The arguments he uses do not apply to the normal situation we face in hospital wards or in the practices of primary care physicians. "The majority of decision-making situations in hospitals and clinics are quite ordinary, and thus do not require emergency procedures."[161]

A different situation emerges in situations with incompetent patients. In these cases it is always necessary that someone other than the incompetent person decides on behalf of the patient. By definition, the incompetent person has already lost his or her capacity to act autonomously, therefore autonomy cannot be violated through a physician's action. The decision makers might be close relatives, legal guardians, or the physician (after consultation with independent colleagues who are not otherwise involved in the case).

The question of a waiver might also be considered relatively uncomplicated. If we are to take the notion of autonomy seriously, it must remain in the power of an autonomous person to put decisions pertaining to the course of treatment entirely in the hands of the caring physician (this power must also entail the right to end this agreement or to make changes to it).[162] A good example of relinquishing powers (in regard to the information patients can reasonably expect to be given) occurs in the case of AIDS, in the context of the levels of CD4+ and CD8 lymphocytes circulating in the blood. These cells are assumed to trigger the defense reaction of our immune system against invading viruses. The steady decline of the health of people with AIDS correlates in a specific way with a corresponding steady decline in the CD4+/CD8 ratios.

Many patients refuse to receive any information about their CD4+/CD8 levels because they find this information too distressing. Others ask to be informed about their immune status only when it bears some relevance to their course of treatment. For example, if their lymphocyte levels fall to a level where it is advisable to start regularly using prophylactics against *Pneumocystis carinii* pneumonia, then they want to be informed. However, they do not want to know anything about the ongoing decline until the next decisions about possible avenues of treatments need to be made. In the view of these patients, which I find reasonable, it is useless to have other information given to them because such information is not tied to any treatment decisions that need to be made and because it causes significant distress to many of them.

The patients who do not want this information might reason that there is not much they can do to influence the continuous decline of their health, and they want to leave it to their physicians to make the right decision. They might have decided that it is not worth going through the psychological trauma to see the rapid and seemingly unavoidable decline of their immunological defense mechanisms at all, especially given that the possible treatment alternatives leave them (at the time of this writing) still with one inevitable endpoint—death. The decision of patients with AIDS to release the physician from his or her duty to inform them about these matters has to be respected. A violation of this patient wish constitutes a violation of the patient's autonomy.

The last of Dworkin's list of justifications for paternalistic behavior seems to me to be problematic. He argues that a physician has no obligation to keep a given patient fully informed if he or she believes that this would be harmful to the patient. Dworkin is not the only one holding this view. David Archard has expressed a similar opinion. He argues that, "If I believe you are doing the wrong thing I should try to get you to do what I think," but that there are "good reasons why I should not coerce or lie to you."[163] He contends that in our daily lives we "do not lie, deceive or coerce," but "we do withhold information and resources we might normally have been expected to supply."[164]

In general, I find this contention doubtful. I do not deceive, lie, or coerce, and it is extremely doubtful that I, under normal (i.e., non-emergency) circumstances, withhold information I "might normally have been expected to supply" because I am aware that the consequence of this omission is identical with any kind of lie or deceiving. Dworkin mentions some reasons for such withholding of information, among them are causing depressions, stress, anxiety, and a general deterioration of health.

In a non–Anglo-American context, however, Dworkin's line of reasoning has been considered sufficient to justify the omission of crucial information, as Jamie Feldman was able to monitor in an

AIDS ward of a large hospital in Paris.[165] This is clearly a case of strong paternalism, where this deprivation of information takes place without prior consent of the patient. In any situation other than those mentioned in the third justification, the omission of certain information, as well as outright untruths on the side of the physician, constitute grave violations of patient autonomy and are unacceptable.

These omissions or lies prevent the patient from making his or her own choices in cases where decisions need to be made that are crucial for the further direction of the patient's life. Such patients are essentially denied the chance to live their own lives. They cannot utilize their second-order capacity (maybe one last time in their lives) in order to go in a different direction, which might be to remain hospitalized, to decide to be treated at home, or, indeed, to refuse any treatment at all. Clearly this would conflict with Millian approaches, where liberty has a very high value, and only certain types of weak paternalism (in cases where patients are not fully informed or are otherwise incompetent) could legitimate a state's interfering with individual life concepts, plans, and actions.

A possible solution to this problem (in which the prevention of harm to patients conflicts with respect for patient autonomy) could be to ask the patient before the diagnostic procedure begins to accept a less than complete disclosure of information about his or her health status, if the full disclosure can reasonably be expected to cause harm. If the patient objects to this proposal and wishes to be fully informed, then a conflict exists between the autonomous person's decision, on the one hand, and what the physician considers to be prudent, on the other. Dworkin argues clearly that physicians do not have to inform patients completely about the state of their health when this could lead to depressions and/or stress. This, however, is normally the case when a patient suffers from a life-threatening or terminal disease. To declare such person—admittedly out of well-intentioned paternalistic motives—essentially incompetent does not seem to be a viable option. This is because, as I have pointed out earlier, it would deny patients the opportunity to make important choices about how they wish to lead their lives in the remaining time left to them.

The nondisclosure of complete information in cases of life-threatening or terminal illness is unjustified for another reason; there is significant danger that the general population could lose their trust in the medical profession should such a practice take place or be considered acceptable. Such anxiety on the side of the population is certainly understandable and, indeed, justifiable. After all, if this kind of paternalism is accepted, how could patients who have apparently minor health problems know for sure that they are not really suffering from a terminal disease and that the physician who presented them with their diagnosis has not deceived them for paternalistic reasons?

Even if we accepted the claim that nondisclosure of full information, or even active misinformation in some cases, is better for the patient's well-being than telling the truth, once we consider the consequences of such behavior for the general physician-patient relationship, it seems beyond doubt that more harm than good would occur overall. Historically, Worthington Hooker, a rule-utilitarian who was an important figure in biomedical research and medical ethics, supported this point of view as early as the nineteenth century. He argued:

> The good, which may be done by deception in a *few* cases, is almost as nothing, compared with the evil which it does in *many*, when the prospect of its doing good was just as promising as it was in those in which it succeeded. And when we add to this the evil, which would result from a *general* adoption of a system of deception, the importance of a strict adherence to the truth in our intercourse with the sick, even on the ground of expediency, becomes incalculably great.[166]

Dworkin seems to agree with Hooker's practical conclusion:

> Given the possibility of explicit waiver, given the state of dependency to which illness reduces us in any case, given the erosion of trust produced by such an exception, and given the difficulty of effective restraints on the use of such powers, the resulting loss of autonomy created by the wide exception of therapeutic privilege is too great to secure general agreement among all those concerned.[167]

Beauchamp and Childress also support this point of view in their standard work *Principles of Biomedical Ethics*. They concede that there might indeed be some situations in which it is in the best interests of the patient not to be fully informed of the risks, for example, of an imminent operation. They believe, however, that there is a great danger of this paternalistic motive being used to justify such omissions (or outright lies) in the day-to-day work of physicians in a clinical context. Thus, they propose something I consider reasonable. In Beauchamp's and Childress' opinion, it is sufficient to warn patients (for instance, before an operation) of the possibility that things might go wrong, in addition to offering them a detailed list of all possible risks. According to this scheme, only if the patient explicitly requests to see this list, should they go ahead and explain the potential negative outcomes.

Beauchamp and Childress quote a group of anesthesiologists to demonstrate that their proposal is also practically operational: "Tell all patients that there are serious, although remote, risks of anesthesia, but . . . allow the individual patient to decide how much additional information he or she wishes to obtain about these risks."[168] Indeed, as the medical anthropologist Jamie Feldman reports, once such a policy is firmly established and patients are given all the information they require, they gain more confidence and trust in their physicians. Feldman describes how patients in a Paris clinic were very well informed at the beginning of their treatment. They were given an in-depth explanation of everything that was known about their disease. However, after a while, both physicians and patients began to realize that the amount of information provided quickly became unmanageable. Patients and doctors began to seek a more traditional, paternalistic kind of relationship, in which the doctor is expected to perform the technical job, doing whatever is in the best interest of his or her patient.[169]

If patients decide voluntarily to forego the option of getting all of the available, perhaps complex, technical information that their physician has mentioned, I cannot think of an ethical reason why the clinic should not respect this wish. If, however, this decision is the result of an inappropriate presentation of the information, perhaps one in which the physician has deliberately buried the patient under preliminary technical information the patient could not possibly understand and evaluate, then there is obvious cause for concern.

If it was up to Gerald Dworkin to answer the question of whether or not patients with terminal diseases should be allowed to buy and use experimental drugs, it seems he would have to look at it from two angles. If the patient with AIDS was the only victim of the potential harmful and counterproductive effect of an experimental agent, then Dworkin's theory of autonomy allows no ethical reason imaginable that would support the conclusion that it is justifiable to interfere with the risk taking of such a patient. His argument against slavery, however, indicates that he would come to a different conclusion on a practical level. The reason for this is that it is hardly ever the risk-taking individual alone who might suffer from the relevant act. Usually friends, colleagues, medical staff in hospitals, and, through public funding of the health care system from tax moneys, even the average taxpayer who has never seen our patient are affected. Hence, Dworkin could argue that certain actions involve harm to parties other than the risk-taking actor.

The question is, of course, whether or not this harm is significant enough to interfere with the individual liberties of an autonomous being. The answer to this question is far from self-evident. In certain cases (such as hang gliding, rock climbing, etc.), we do allow persons to take significant risks just for the sake of having fun. Hence, it seems unjust to interfere with the person who is struggling to survive and is willing to take significant risks, while we do not interfere with the actions of the perfectly healthy person who is taking risks of a similar magnitude and significance for the mere fun of it. (These scenarios are analogous in terms of the likelihood of the relevant risk taking leading to burdens for third parties.) Moreover, it does not seem appropriate to use minimal (in many cases even negligible) negative consequences for society in general to prevent the satisfaction of vital (survival) interests of terminally ill persons.

WEAK PATERNALISM: TOM L. BEAUCHAMP, JAMES F. CHILDRESS, AND RUTH R. FADEN

The American bioethicists Tom L. Beauchamp, James F. Childress, and Ruth R. Faden have advanced their theory of autonomy in two major works of modern biomedical ethics.[170] They interpret individual autonomy as an extension of political self-rule or self-governance by the individual. This excludes both controlling interferences by others and personal limitations (such as inadequate understanding) that prevent meaningful choices. Similar to Dworkin and Young, Beauchamp and Childress also claim that the "autonomous individual freely acts in accordance with a self-chosen plan, analagous to the way an independent government manages its territories and sets its policies."[171] They characterize an act as autonomous, if and only if, X acts intentionally, with understanding, and without controlling influences.[172] They propose to judge an act not on the basis of the unrealistic postulate of complete autonomy, but on the basis of whether or not it is substantially autonomous. The reasoning behind this is persuasive, since in the real world people are never fully autonomous.[173]

Faden and Beauchamp present us with an interpretation of the autonomy of acts as existing on a continuum that extends from completely autonomous acts (the undoubtedly ideal situation that can barely ever, if at all, be reached under real world conditions), over a gray zone of substantially autonomous acts, to the other extreme of the completely nonautonomous act. Their description of a continuum is convincing, especially when we recall their three criteria of intentionality, understanding, and noncontrolling influences. "There are *both* different conditions under which and different degrees in which autonomous action may be enhanced or compromised."[174] However, it is important to emphasize that each of the three necessary conditions must be, to a certain degree, satisfied before we can accept an act as substantially autonomous. Beauchamp and Childress claim that if only one of these conditions is unsatisfied, the resulting act is nonautonomous.

The condition of intentionality requires that an act be carried out in "accordance with a plan."[175] In order to exercise such an act in the first place, I must have an understanding of its implications (risks, consequences, etc.). Faden and Beauchamp argue that this understanding need not reach the level of omniscience, but only an understanding of the foreseeable present and future consequences. The obvious difficulty in our case is that, when we take experimental drugs, we cannot, by definition, foresee the consequences. This being the case, in order to make a voluntary and autonomous decision, it would be enough for us to be aware of our limited knowledge and to be willing to take the risk. As I have already pointed out, the same would be true for many other decisions, such as the decision to climb an alpine peak by a new route in winter or to choose a sport with similar risk potentials. The only thing we can know for sure is that there is a risk of an unknown magnitude to be harmed. We know also that this harm ranges from minor injuries, to irreversible spinal injuries (which might leave us severely disabled for the rest of our lives), to the possibility of accidental death.

The autonomous decision to take unapproved medical drugs creates an analogous situation, in which people with AIDS have a general understanding of the risks involved in the application of these substances. They do not, however, have any knowledge about the exact nature of the potential (side) effects of a certain drug. Against this view, Ten has argued that:

> if all that is required is that people know that they are ignorant about drugs, then indeed we can attribute such metaknowledge to them. But metaknowledge of this kind is compatible with first-level ignorance, which cancels voluntariness in risk taking. Thus an ignorant drug user might be unaware of the high risk that she is taking because she does not know that the drug can cause very great harm. She does not therefore voluntarily take the risk of grave harm.[176]

In a persuasive reply to Ten's objection, Häyry argues that a truly nonautonomous decision would exist in this case only if the decision maker lacked "both the first-level knowledge and the metaknowledge concerning her own ignorance."[177] Of course, the more persons use a certain drug, the better the general first-level knowledge will be, assuming that the experiences individuals have will be shared with other patients.[178]

I hope to have made clear that I believe substantial metaknowledge about the nature of the risks a person is taking when he or she uses experimental drugs is sufficient for an autonomous decision to take experimental drugs. To better substantiate this position, however, I think it appropriate to add the following remarks on the matter of the quality of the existing first-level knowledge. Before I do so, however, it should be mentioned that more than a decade after the first AIDS cases were diagnosed, no studies have been undertaken to gather information about the quality of the first-level knowledge of persons who use experimental agents. This is surprising, given that it is known, even among AIDS experts, that the use of experimental agents among people with AIDS is fairly widespread. Still, as this is the case, I will have to rely upon my reading of information that such patients have made available in their informal newsletters, personal communications, and other such sources of information.

People with AIDS, such as Jon Greenberg, a self-identified "alternative treatment activist," are aware of the problems surrounding the lack of knowledge about experimental agents. He wrote:

> Some dismiss alternative treatments, regardless of the evidence demonstrating efficacy, and others defend alternative treatments, regardless of evidence demonstrating toxicity or lack of efficacy. The reality of most alternative therapies lies somewhere between these two extremes. Some alternative therapies may be effective, some are clearly ineffective, and most posses some degree of toxicity. The chief difficulty with using alternative therapies is a lack of empirical data and an absence of scientific interest in these compounds. Presently there is no research infrastructure to address systematically the potential benefits and risks of alternative treatments.[179]

He warns that "since toxicity studies on most alternative therapies have not been conducted—and since many alternative treatment practitioners often recommend very high doses—it must be asserted that they may be toxic."[180] Kevin Armington warns his fellow people with AIDS "to beware of claims that a treatment is a 'cure' or will have miraculous effects." He notes that "prohibitively high prices for experimental treatments can be the tip-off to potential fraud."[181] An anonymous article written by a person with AIDS warns, "many of the fraudulent treatments in the community have 'secret' compounds. . . . There are actually 'secrct' compounds out there that are no more than mothball and water or creosote [the black tar residue on railroad pilings]."[182]

Another strategy proposed by Armington is to "not rely on only one source to make your decision."[181] Consistent with this suggestion, AIDS service organizations have begun to compile information about alternative treatments in systematic ways in the form of information files on different drugs and also in the form of whole studies. For instance, the Deutsche AIDS-Hilfe, an umbrella organization of German AIDS organizations, which is largely sponsored by governmental funds, has published a book that gives an overview about what the authors call "unconventional medicine."[183] It is the result of a research project undertaken by AIDS activists, medical researchers working at the Free University of Berlin, and general practitioners (GPs) who treat AIDS patients and have some experience with the alternative treatments mentioned. (The study is available free of charge to interested people.)

In the Minnesota *PWALive* magazine, Carlton Hogan wrote a series of articles under the heading "The Prudent Pariah," which gave readers basic information about technical medical terms and the functioning of the immune system. Hogan explains that his motive for writing these articles was to help people with AIDS make better-informed decisions about their own care. His intention is to increase access to valuable sources of information. Hogan realizes that he cannot supply "the answers," as he put it. "I have no idea what works to keep AIDS at bay, and my particular regimen might kill other people. What I would like to share are the tools you can use to make those incredibly personal decisions for yourself."[184]

Hogan, Armington, and Greenberg are just three examples of people with AIDS advocating experimental therapies. As such, they may serve as counterexamples to the claims of theorists such as Shorr and Annas, both of whom have characterized people with AIDS as easy to manipulate into the use of experimental therapies because of the threat AIDS poses to both their physical and their psychological well-being.[185] As a matter of fact, many people with AIDS, in addition to the organizations supporting their wishes to use experimental drugs, are not only very well informed on a meta-knowledge level about the risks they run, but they also try to gather as much first-level information as possible. As an article written by an AIDS patient, published in the New York City PWAC (People with AIDS Coalition) *Newsline*, correctly noted: "You are ultimately responsible for what treatment you choose to use and this responsibility should not be taken lightly. Arm yourself with knowledge, patience, and never abandon your common sense."[186]

As Nussbaum's authoritative account reports, first-level knowledge actually helped several thousand people with AIDS, from 1986 onward, to survive because they used an efficient prophylactic for *Pneumocystis carinii* pneumonia.[187] This type of pneumonia was the leading cause of death in the early days of the AIDS epidemic. However, aerosolized pentamidine, which was the prophylactic these people used, was at that time not approved by the U.S. FDA. Consequently, it had the status of an experimental drug.[188] The final FDA approval came three years later. In the meantime, however, word-of-mouth propaganda and efficient networking of primary care physicians saved thousands of lives that would have almost certainly been lost had they waited for this particular drug to be approved.

This kind of networking among people with AIDS is also used in attempts to gather information, as many posters in the usenet newsgroup science.medicine.aids demonstrate. The following quote is just one example among innumerable others. George M. Carter, a person with AIDS living in New York City, posted (on April 15, 1995) a note to this newsgroup that contained and requested information regarding an experimental agent. He wrote: "I am writing to inform and query knowledge upon some information regarding yet another remedy of fungal origin, to wit, the off-puttingly named Stinkhorn."[189]

Carter then went on to provide the information that he was able to gather about this remedy. He concluded his poster by writing: "My question is: has anyone had experience with using Stinkhorn? I think it is important to investigate possibilities, while remaining skeptical."[189] In subsequent days, other people with AIDS, as well as physicians, contributed their information to the newsgroup and so improved the overall level of information available to interested subscribers. Other support groups, such as e-mail lists dedicated to the sole purpose of disseminating treatment information among people with AIDS, have been established in the meantime. A similar story has been told about an antiviral drug, ribavirin, which was smuggled over the Mexican border by Jim Corti. A physician advised him that this drug might help and would certainly not kill because, as the physician pointed out somewhat bluntly, "Ribavirin's been swallowed by thousands of Mexicans and Europeans. Nobody seems to have keeled over from taking it."[190]

The last example I provide concerns a debate in the U.S. Congress and Senate over whether vitamins and nutritional supplements should lose their food status and subsequently become prescriptive drugs. The motive for this move basically concerns outrageous, and grossly misleading, claims about the positive effects of these supplements for both general health and the particular health of persons with immune disorders. In this context, alternative treatment activists have again shown that they are neither as easily manipulable as certain bioethicists have claimed, nor the easy targets that common sense might have us believe. For instance, Derek Hodel, of the Gay Men's Health Crisis, pointed out that there are problems with fraudulent health claims made by producers and marketers of dietary supplements. However, he realizes also that "people living with HIV disease, faced with narrow enough (and mostly inadequate) treatment options from so-called mainstream providers, are understandably concerned with the prospect that their options might be further diminished."[191]

However, we have to keep in mind that a reasonable number of options are only one side of the coin. Spencer Cox of the influential New York City–based Treatment Action Group argued that all too often bioethicists and political AIDS activists make policy on the basis of an exceptionally well-informed group of patients. Cox is rightly concerned that AIDS activists have made treatment policies based on their own experiences, i.e., that of a few well-educated patients with a lot of leisure time. Clearly the majority of PWAs, both in the Western world and in developing countries, do not fall into this category.[192]

He correctly notes that there are significant differences between the means of educated, white, middle-class people with AIDS to access and gather knowledge about treatment options and those of people from relatively marginalized groups of the population. This practical problem needs to be addressed urgently, not just in the United States, but also in Australia, where IDUs and people from non–English-speaking backgrounds remain both disadvantaged and underrepresented in AIDS organizations.

Having acknowledged these problems, however, it is still necessary to stress that it is in a given patient's best interest to have access to, and a means of evaluation of, the available first-level knowledge about a certain experimental agent (or alternative treatment options for a certain condition). Clearly, this is only possible insofar as there is first-level information available about a certain experimental agent. It is important to remember that, especially in cases of experimental agents outside of clinical trials (where it is likely that no preclinical trials have been undertaken), it is possible that there is no significant amount of first-level information available. This clearly makes the decision of a person with AIDS to use such an agent more risky than to decide to join a clinical trial of an experimental agent that has already been tested in preclinical trials.[193]

In practice, however, there are some indications that the situation of a person who joins a standard clinical trial is not that much different insofar as taking the chance of an unknown risk is concerned. One partner of a person with AIDS mentioned to me that the patient was required to sign an informed consent form, which made explicit that the risks of the study were not well known.[194] The difference in the risk magnitude is only due to the fact that preclinical trials have given some indications about possible toxicities and that the patient is under permanent observation by medical staff. Moreover, it seems that the influences to which the patients are subjected are overwhelmingly those of principal investigators, their physicians, community-based AIDS organizations, and other public health officials, who advise them, in the vast majority of cases, not to take unapproved drugs. Therefore, the probability that controlling influences have manipulated people with AIDS-defining diseases to take unapproved drugs appears to be minimal. Indeed, in reality, it appears that the reverse situation is far more likely to occur, whereby patients are urged to join clinical trials and to avoid unapproved remedies (even as supplements).[195]

The examples just presented indicate that the alleged peer pressure of alternative treatment groups and activists upon people with AIDS-defining diseases can, with some degree of confidence, be discounted. Still, the fact remains that there may well be a small number of PWAs who are manipulated by clever "quacks" or through peer pressure (which includes their lovers and close friends). It is for this reason that I think it is always necessary for a particular person with AIDS to prove that he or she has substantial metaknowledge of the risk he or she is intending to take.

Thus far, I have presented a number of constructions of autonomy and analyzed their implications for people with AIDS-defining diseases who wish to take experimental drugs. At this stage, I think it is important to emphasize that none of the previously discussed concepts of autonomy are consistent with the prevention of access to such drugs, at least given the satisfaction of the following two conditions: first, the patients in question are the only ones who will be affected by the potential negative consequences of their choices, and second, they can demonstrate sufficient metaknowledge of the risk(s) entailed.

Both Beauchamp and Childress argue, however, that autonomy is only one value among others. They discuss at some length the principles of nonmaleficence, beneficence, and justice. Paternalistic beneficence is, in their view, the strongest competitor against an acceptance of individual autonomy as an overriding principle. Paternalism is interpreted in this theory as "the overriding of a person's wishes or intentional actions for beneficent reasons. . . . Our thesis is that beneficence alone is the justification of paternalism."[196] Their claim that strong paternalism is the only controversial form is certainly well accepted in the bioethical debate over paternalistic interventions. Far from uncontroversial, however, are their views about both the situations in and the circumstances under which strong paternalistic interventions are ethically justifiable. Strong paternalistic interventions in individuals' lives are justifiable in their opinion only if:

1. A patient is at risk of a significant, preventable harm.
2. The paternalistic action will probably prevent the harm.
3. The projected benefits to the patient of the paternalistic action outweigh its risk to the patient.
4. The least autonomy-restrictive alternative that will secure the benefits and reduce the risks is adopted."[197]

These conditions are somewhat tricky because Beauchamp and Childress, although claiming that condition 5 is crucial, do not give us many indications about how to weigh the conditions against each other if only some of them are satisfied. It also remains unclear whether each of them is only a necessary condition and whether their combination makes them a sufficient condition. I assume that the latter is the case.

Take, for example, an AIDS patient who has the urgent wish to buy and apply drug X, which is supposed to be successful against an opportunistic infection of the lungs that will kill him or her very soon if a successful intervention is not made. In this context, "supposed to" means based on information that the patient has retrieved from newsletters of fellow AIDS patients and AIDS service organizations and from hearsay stories of sick friends. He or she is well aware that the risks involved in taking drug X cannot be estimated. In particular, the patient has no knowledge about the long-term effects of this drug, yet also knows that not taking the risk will leave perhaps six months before death, most which will not be worth living due to the effects of chemotherapy.

Now, clearly, our patient satisfies condition 1, as he or she is at risk of injury (the drug might do more harm than good). Condition 2 provides an argument against intervention, as the risks of the intervention are significant, given that our patient would die without a "miracle drug." Preventing the patient from buying and applying the drug will not save this person's life (and, therefore, in accordance with Young, his or her long-term dispositional autonomy). Condition 3 is an empirical question. It is unclear how big the risks are, though the patient is well aware of this and willing to take the chances.

Condition 4 can easily be met on the side of the paternalist. What does "feasible and acceptable" mean? To whom? The patient, or the physician who tries to prevent the patient from doing what he/she wishes to do? What are the criteria for feasibility and acceptability? Is this not merely a vague phrase, which will be conducively interpreted by the strong paternalist who wishes to intervene? Who decides which infringements are ultimately acceptable? Certainly not the patient who considers unacceptable any intervention that will prevent doing what he or she wishes in this situation.

The "crucial" fifth condition adds another point on the side of the person with AIDS, as the infringement of the respect for autonomy principle is enormous.[198] If I follow Beauchamp and Childress's own practical interpretation of these criteria, the patient dealing with AIDS in my example would have his way. They argue that, in the case of a Jehovah's Witness who refuses a blood transfusion, a vital autonomy interest is at stake, and strong paternalism would therefore be unjustifiable. In the next section I will analyze a strong paternalistic approach that shows why an intervention in plans, such as the one our hypothetical AIDS patient has made, violates vital autonomy interests.

The final condition of Beauchamp and Childress is also open to debate and dispute. What constitutes "the least infringement necessary in the circumstances" is generally open to a wide range of conflicting interpretations. In our specific circumstances, however, the least infringement necessary would still be a state of permanent intervention, as I shall subsequently demonstrate in my discussion of Robert Young's philosophical justification of strong paternalistic actions. Such an infringement again violates condition 5, and therefore, Beauchamp and Childress would have to support our patient's view. Clearly, their approach supports the general idea that we have an obligation to not override the decisions of autonomously acting individuals, for strong paternalistic reasons, unless the violation of autonomy is minimal while the weight of beneficence is truly tremendous. The outcome of this balancing of harms and benefits is ultimately dependent upon the actual situation. In accordance with this theory, people with AIDS would not be prevented from buying and applying experimental drugs on either weak paternalistic or strong paternalistic grounds.

STRONG PATERNALISM: ROBERT YOUNG

The Australian philosopher Robert Young interprets individual autonomy as a character ideal or a virtue (among other character ideals) and considers it intrinsically valuable. He agrees, however, that the interpretation of individual autonomy as a character ideal does logically not exclude the possibility of autonomy also being of instrumental value. Young emphasizes, "To value autonomy intrinsically is not incompatible with valuing it instrumentally."[199] He considers autonomy as intrinsically valuable "because it *provides* the foundation of moral personhood and *gives us* dignity and standing as moral agents."[200] Earlier in *Personal Autonomy*, however, he writes, "In taking autonomy to be of fundamental significance for our understanding of moral personhood, I have taken it to be an object of desire *whose primary value is not that of a means to some further good.*"[201]

Assuming that autonomy fulfills the two functions of providing the basis of moral personhood and giving us dignity and standing as moral agents, then it becomes fairly obvious that in both cases autonomy has only instrumental value. Still, I will not discuss whether Young's analysis should be accepted or rejected because Young does not ascribe absolute value to the autonomy to which he has attributed intrinsic value. Given that he puts autonomy at the level of just one (important, intrinsically valuable) character ideal among a number of other ideals, autonomy has no overriding importance and must compete with other values. Hence, we are left with autonomy being on an instrumental level as described by Mill. Here too, autonomy remains in permanent competition with other intrinsic and instrumental values.

Young agrees with Harry Frankfurt and Gerald Dworkin that authenticity is not only a prerequisite of autonomy but that "the authentic person is the autonomous person,"[202] in a technical sense. Young's theory of autonomy is firmly based on the Kantian ideal of the autonomous individual possessing his or her own will. This will enables the person to govern his or her own way of life. Fortunately, in contrast to Kant, Young does not subscribe to the idea that the values on which we act should be necessarily rational or even of universal validity. For Young, it is only crucial that these values are our own values and that our acts are results of a life plan that is compatible with these values. Young's rejection of such universal validity is, in my view, highly plausible. Individual interests, which are the results of my personal, subjective life concept, are not necessarily of universal validity, and it is not clear why they should be. The same is true of my feelings, needs, desires, and values, which form the basis of my life plan.

Freedom is a necessary, but not a sufficient, condition of individual autonomy. Just as Feinberg and Frankfurt do, Young recognizes that insufficient resources, such as poverty, restrict options and, thus, autonomy.[203] This seemingly trivial point is of some importance for the evaluation of the situation of AIDS patients in countries without universal health care. For instance, in the United States, in addition to gay men, it is predominantly people from ethnic minority backgrounds who are affected by AIDS.[204] In particular, poverty, and the lower educational level with which it is often intertwined, is the primary factor that results in a significant practical limitation of the options for these people. I will discuss the problems related to this subject matter in greater detail in Chapter 3.

Which dispositional capacities does an individual need to be considered an autonomous person? At the beginning of his book, Young offers a description of such an individual: "those who themselves determine what they will decide and do, rather than have these dictated to them by circumstances or by others, are autonomous."[205] Still, it may be asked what implications this criterion has. Young says that the aforementioned self-directedness needs to be filled with content. He claims that judgments about autonomy are primarily judgments about dispositional autonomy. Interpreted in this sense, they are judgments related to the question of whether a given person was able to convince others that his or her intended act followed an overall life plan or life concept.[206] He goes so far as to claim that the degree to which a person's life is autonomous depends on how completely this individual is following a life plan that explicates his or her choices. Indeed, to develop or to rethink one's life plan, something important is needed—reason.

Theoretically, Young's postulate is very plausible. Practically, though, there are significant problems, some of which might even be strong enough to render this criterion impossible to implement. For instance, we would need to appoint authorities who have the capacity to determine in situations of conflict whether someone's intention to act in a certain way derives from a life plan or whether it is, in a Kantian sense, heteronomous. In such cases, the decisions made would likely oppose an intended act when the act appears risky or when the judges deem the plan imprudent, or simply wrong. Such an occurrence would be predictable under circumstances of extreme psychological stress for the individual in question, in which changes in one's life plan are irreconcilable with earlier life concepts.

Youngian guardians would have to conclude that, in such situations, the intended act is not consistent with the known (earlier) life plan(s) and, hence, declare the person nonautonomous in regard to the planned act. In such a society, a person who is facing a terminal disease and makes changes to his or her life plan would have the additional burden of convincing paternalistic guardians that actions are consistent with a newly adopted life plan. This will undoubtedly be impossible for many of these people, even if their plans are otherwise carefully thought out. This fact, however, does not allow the conclusion that therefore they have no coherent life plan. It only shows that they are unable to convince the relevant institution that, despite appearances that indicate otherwise, they really do act on the basis of such a plan.

Young is undoubtedly right when he argues that some people might use dispositional autonomy for imprudent purposes. It is ill-used in such cases, and it would probably have been better if people had not exercised their autonomy in that way. In such cases, Young proposes to restrict the act-autonomy of the individual in question. Such a restriction would occur not only where others will be affected by consequences of the ill use of the autonomy, but also when the acting individual is harmed. As we have already seen, respect for individual autonomy has no absolute value ascribed to it in Young's theory; it is to be seen relative to other values.

In his book, Young tries to specify circumstances under which it is justifiable to restrict individual autonomy. He rejects Mill's proposal and those of Mill's disciples because their position "is untenable in view of the fact that strong paternalism is sometimes needed to preserve autonomy. Those who seriously value autonomy cannot remain content with weak paternalism."[207] Young defends strong paternalistic interferences as morally justifiable in all those cases where the intervention is the only reasonable means of preventing an individual from risking long-term dispositional autonomy.[208] This is so because "to be autonomous is to be one's own man or woman; it is not merely to be free but to be free to order one's life in a unified way according to a plan or conception which fully expresses one's own individual preferences, interests and so on."[209] If people intend to act in a way that indicates that they are willing to risk this long-term self-determination, then strong paternalistic acts are not only justifiable, but morally necessary, if we value individual autonomy, argues Young. This individual long-term self-determination is at stake when people intend to sell themselves into slavery, and this is it what makes such intentions and such acts wrong and presents us with a motive to intervene. Since we should try to protect the long-term dispositional autonomy of such persons, we need to prevent them from selling themselves into slavery.

Young insists on the difference between advocating a strong paternalistic action in order to "protect an individual from harming his [or her] own interest," on the one hand, and his strict rejection of legal moralism on the other, which is intended to "defend the individual from one alleged sort of harm."[210] The motive for intervention that Young would accept in the case of slavery, then, is not the abhorrence most people in society have when they think of the situation of a person who has to live as a slave, but the threat to the long-term dispositional autonomy of the slave.

Young tries to deflect criticism that in his theory the strong paternalist assumes de facto the role of the slave owner each time he or she prevents an autonomous individual from doing what that person wants to do. He replies by distinguishing between the goal of the ultimate preservation of long-term dispositional autonomy and the strong paternalistic violation of actual or occurrent autonomy. According to Young, only the latter may be violated in order to preserve the former.[211] This view implies, of course, that the autonomous individual is not entitled to make at least one decision, which is to risk long-term dispositional autonomy. Ironically, this means that Young's premise contains what his argument tries to prove: it is ethically justifiable to sacrifice occurrent autonomy for the sake of preserving long-term dispositional autonomy. Therefore, Young, contrary to his own judgment, fails in his attempt to confront Feinberg's objection to strong paternalism.[212] Feinberg has argued that "the principle that shuts and locks the door leading to strong paternalism is that every man has a human right to 'voluntarily dispose of his own lot in life' whatever the effect on his own net balance of benefits [including freedom] and harms."[213] Let us now look at AIDS further to help illustrate Young's difficulties.

Certainly, people with AIDS diseases who intend to use untested experimental drugs take a risk that entails the distinct possibility of sacrificing their own long-term dispositional autonomy. It seems beyond doubt, therefore, that a strong paternalist of the Youngian creed would have to intervene in order to preserve these persons' dispositional autonomy. A deeper reflection on the specifics of this case, however, calls into question the self-evidence of this conclusion. The dispositional autonomy of the average patient with AIDS-defining diseases differs in some significant aspects from that of most other persons without terminal diseases. The endpoint of these diseases, in the absence of any successful standard treatment, is death, which implies the permanent loss of current as well as dispositional autonomy.

More than ten years after the first people with AIDS-defining diseases have been diagnosed, there is no difference in the endpoints of persons using experimental drugs, those joining clinical trials, those who choose to do nothing, and those who use approved drugs. The vast majority die. The strong paternalist would necessarily fail to "increase the dispositional sovereignty of the agent," a condition Young requires be met in order to make a strong paternalistic intervention into the life plans of autonomous agents morally acceptable.[214]

Indeed, John Dixon who is, overall, against access to experimental drugs for people suffering terminal diseases, even argued that such patients have good reasons to justify their claims to greater control over the treatments. He pointed out that:

> a catastrophic situation undercuts the general case for medical paternalism, because established medicine is simply without the means to guarantee that if it thwarts catastrophic patients' therapeutic wishes, its doing so will result in "their own good." A catastrophic illness has a leveling influence on the relative claims to therapeutic competence of the physician and the patient, and this justifies such patients' claims to a greater measure of self-determination in their treatment.[215]

The strong paternalist who wishes to prevent people with AIDS-defining diseases from taking experimental drugs would essentially have to permanently restrict the autonomy of these patients until they finally die of this illness. If the paternalist did so for only a limited time, then the people with AIDS who initially intended to buy and apply experimental drugs, would most probably attempt to do so again. This is at odds with Young's idea that "paternalistic intervention should not be directed at producing a continuing state of dependence, but rather be used to increase the dispositional sovereignty of the agent."[216] A patient with full-blown AIDS has an average life expectancy of one and a half to three years. This patient's individual sovereignty cannot be enhanced by preventing him or her from buying and using experimental drugs during this time. This is not to claim that people with AIDS-defining diseases cannot harm themselves significantly if they buy and apply the wrong drugs. But, to quote Young again (applying his argument in favor of voluntary euthanasia to the situation of people with AIDS-defining diseases), "in acceding to their request . . . we preclude no significant future choices."[217]

We do not preclude significant future choices because, even if these AIDS patients adhered to standard treatment procedures, they would die. I would consider a significant choice a choice that would allow them to preserve their long-term dispositional autonomy. An indication that Young favors this conclusion can perhaps be seen in his practical support of voluntary euthanasia.[218] In this context, he argues that we should respect the autonomous decision of terminally ill patients to die because there is no long-term autonomy that can be preserved.

It is important to emphasize that my argument does not apply in at least two types of situations. The first is a context in which successful standard treatments are available, while the second occurs when patients with life-threatening diseases insist (for whatever reason) upon using remedies that are proven to be useless, or even harmful. These situations, however, are not currently faced by people with AIDS.

The Youngian strong paternalist could still insist that those patients risk losing their dispositional autonomy at an earlier stage than those who use the already existing, but largely unsuccessful, standard treatments and this possibility provides the ethical basis for an intervention. However, even though this possibility is not entirely unrealistic, what would the alternative look like as offered by our strong paternalist? He would have to eliminate the dispositional autonomy of people with AIDS for the (short) rest of their lives.[219] Therefore, people with AIDS who are subjected to strong paternalistic guardians would be worse off because they would forfeit their presumably last chance to determine by and for themselves how they want to live their last months and, equally important, how they wish to fight the disease that threatens to kill them prematurely. Clearly, no matter which way one looks at the AIDS example and the question of strong paternalism, people with AIDS would be worse off in the case of strong paternalistic interventions.

THE IMPORTANCE OF RESPECTING INDIVIDUAL AUTONOMY

Autonomy has been considered by some authors to be only of instrumental value. Others think it has intrinsic value. Still others attribute it both types of value. Whatever decision individuals may reach in this regard, it is clear that in all of these cases, we should have an interest in maximizing individual autonomy. In the case under consideration, this means, at the very least, that an agent needs to have a reasonable level of competence which allows for an understanding that a risk is involved should he or she use experimental drugs. People with AIDS or other terminal diseases are not incompetent per se to make decisions about the use of experimental agents. As I have argued earlier in this chapter, the mere fact that they suffer a terminal illness does not mean that they are consequently incompetent to make decisions related to that particular life-threatening condition. The use of standard therapy is not helping to save these patients' lives. Hence, the strong paternalistic motive proposed by Young, that is, to interfere with occurrent autonomy in order to preserve long-term autonomy does not apply under the particular circumstances of a lethal illness that is leading to the death of the sick person within a relatively short period of time. (As mentioned earlier, this also seems to be why Young supports voluntary euthanasia.)

What implications does this argument have for people with a terminal illness such as AIDS? Patients with full-blown AIDS die, on average, within a few years of diagnosis.[220] I argue that they should be permitted to choose between experimental drugs and standard treatments. It may well be that their use of experimental agents will lead to death; in fact, this is quite likely. But I cannot see that an argument can be found, at least on the basis of paternalistic reasons, that would justify forcing these people to use instead a standard treatment that is also known to fail, even if it offers a short-term benefit.[221]

Ultimately, access to experimental agents increases the choices for people with terminal illness. It increases their real-life options and their autonomy, too. For people with AIDS, the importance of being in control of their lives is perhaps best signified by the strong support for voluntary euthanasia given by AIDS organizations representing patient interests. Bradley Engelman of the Australia-based Victorian AIDS Council, who suffers from AIDS himself, reportedly said, "legalizing euthanasia gave AIDS sufferers a tiny piece of control" over their lives. "The thought of being able to die peacefully and quietly with the people you love at the time of your choice is the one thing that we need to really help us continue."[222]

A possible counterargument to this view could be that, even in these few years, people can actually harm themselves. The drug they choose could hasten their death or result in their dependence upon others. I shall only discuss the paternalistic side of this argument here, and will return to the second half in the other parts of the book. Assume that people harm themselves by taking the unapproved agent X instead of the approved agent Y, which is not saving their lives but is at least not harming them. On the one side, we have the chance of harm to self that is fairly limited because people with full-blown AIDS have only a few years left and their quality of life decreases dramatically. On the other side of the equation, we have a permanent intervention in these patients' lives in order to prevent them from taking a certain drug they wish to test. The patients who believe that this drug might save their lives will become more and more desperate to try the drug simply because they are running out of time.

Therefore, the magnitude of the intervention increases proportionately over time—the closer the patient comes to death—and, in many cases, this will ultimately lead to patients being declared incompetent in order to prevent access to an experimental agent. The price paid for the avoidance of potential harm (which, even though likely, is still far from certain to occur) is a permanent intervention in the life-and-death decisions that a terminally ill patient makes. This price is too high.

Is it really necessary to expect that these people prove to us (physicians, relatives) that they have some sort of a rational life plan, as Young and Dworkin seem to suggest? I consider this premise, which relies on Kantian reasoning, very problematic. Standards of rationality are not universally consistent and are always correlated with other normative systems, such as religion, and perhaps influenced by social factors such as race, class, age, gender, and ethnicity.

Standards of rationality are obviously not transhistorical but always relative to those societal determinants that are actually predominant. The request to be rational is problem-laden insofar as it allows wide-ranging interpretations as to what rational means in a given historical context. Rationality that is interpreted in empirical or scientific terms excludes the numerous interpretations of our environment that constitute the respective life realities of many people. On the other hand, should we decide, for example, to accept Christian religious interpretations of life, or those of Jehovah's Witnesses, in spite of their contradiction of scientific rationality, then I can think of no reason not to accept the interpretations of very small religious sects or, indeed, those held by only one person—our patient.

A picture of rationality that is dominated by a scientific empirical interpretation of the environment would declare basic parts of the life plans of many people invalid. Hence, the decisions they make on the basis of certain beliefs might be overridden, and their attempts to live their own lives would fail because someone with the ultimate legal power to define rationality would declare them nonautonomous. If, however (as I would suggest in our context), rationality means only to be aware of possible consequences of one's actions or that one simply cannot foresee the possible consequences, then there is little to say against such particular, normative premises. I wish to emphasize that in any formulation of rationality, however minimal, we must be careful not to set standards that exclude certain beliefs, for instance those that are religious. (Religious beliefs, I contend, should not be excluded even when they are not based in a religion that is established and popular at a certain point in time.)

Another criterion that we might be tempted to use to identify the rationality of a certain belief is perhaps whether it is logically coherent. Thus, in this case, the belief system must be coherent within its own framework. I would maintain, however, that even this criterion is unnecessary in judging whether certain end-of-life decisions should be respected. If one's belief system is coherent, then that is certainly a plus, but ultimately, an incoherence is an insufficient basis upon which to reject the decision of a terminally ill patient. It is only important whether the patient is aware of the risk encountered should he or she buy and use experimental agents.

Is it necessary that the patients in question be required to prove that their intention to use experimental drugs, or any other treatment-related decision, be coherent with their individual life plans? The postulate of authenticity of value systems and basic convictions as a necessary condition for autonomy is theoretically quite reasonable, and on that level, I have little objection to it. However, when it comes to the practical application of this criterion, I am no longer sure if it really deserves the status of a logically necessary condition. Sooner or later, relatives or the responsible physician (or independent guardians) will have to make practical decisions in regard to the patient's decisions they consider problematic.

At this time, they will decide whether the patient's decision is autonomous or whether it is, in a Kantian sense, heteronomous. Depending on what common sense dictates, the criteria used to make a decision about the authenticity of a given patient's intended action will vary greatly. It may well be that the same decision that would lead to a verdict of incompetence in one society would be accepted without hesitation in another. Ultimately, all that is required and feasible is that patients are aware of the risk that they intend to take. When they have this knowledge, and are aware of the possible alternatives (other alternative treatments, standard therapy), they should be able to make up their minds about their preferred course of action. I do not see any need for them to provide evidence to second or third parties that their decision matches their life plan. In fact, as I have already pointed out, people with terminal illnesses may well change their life plans as a result of being diagnosed with a terminal illness. A new or modified plan might be incomplete or only existent in fragments. Again, this is an insufficient reason not to accept their treatment-related decisions.

As I have mentioned earlier in my discussion of Joel Feinberg's liberal approach, it is important to look closely into the authenticity of a person's act, not only when we think the intended act is not in the person's best interest, but also when we find the intention reasonable. Independently of this requirement, however, it is clear that many people facing death change the direction of their lives a last time and make decisions containing or leading to fundamental directional changes. Perhaps their new insights, even their newly generated value systems, will look strange to the primary care physician and the patient's relatives. This alone might often lead to a verdict that the determinations are the result of a heteronomous decision-making process. I think that, for this reason alone, the criterion of authenticity cannot be applied as a necessary condition for an autonomous act. Clearly, when it is possible to prove that an intended act is authentic, we are provided with another important reason to accept the wish. However, should this turn out to be impossible, then it does not follow that we are entitled to act against the declared wish of the patient for the mere reason that we are unable or unwilling to see that his or her intended acts are autonomous.

In cases such as this, I think it is sufficient that the patient be aware of the potential consequences of his or her act. Should this criterion be impossible to satisfy for practical reasons (there are no known potential consequences, for example), it should suffice to provide evidence that the patient has a substantial understanding of the risk potential of the intended course of action.

SUMMARY OF THE PHILOSOPHICAL DEBATE

In this chapter, I have discussed different weak paternalistic theories of autonomy and one strong paternalistic approach to autonomy. The purpose was to determine how these classical theories answered the question of whether the interest of autonomous people with AIDS-defining diseases in buying and using unapproved, experimental drugs should be respected by society. My argument is deliberately limited to people with a terminal disease such as AIDS, which has no successful standard treatment available to those who suffer from it.

Weak paternalists cannot possibly provide an argument for an intervention to stop people with AIDS-defining diseases from buying and using experimental drugs. In order to counter such proposed intervention, it only has to be certain that a given patient, who wishes to use experimental agents, is aware of one of two types of information. Either the patient must be aware of the (known) risks associated with using a certain experimental agent, or if nothing is known about this agent, the patient must be able to understand that the risks associated with taking this particular drug are unknown.

All of the attempts to support a paternalistic intervention, which utilized arguments such as insufficient information, unknown risk magnitude, and psychological stress related to a life-threatening disease, failed to support the interventionist cause. The proposal to assume that people with AIDS diseases are incompetent was unconvincing. In addition, it violates normal standard procedures that require those who act paternalistically to prove incompetence on a case-by-case basis.[223] Finally, the case was made that even a strong paternalist, such as Robert Young, would be obliged to observe nonintervention in the case of a person with AIDS who wishes to buy and use experimental drugs. This argument drew on Young's own construction of the ethically legitimate reasons for, and the purpose of, strong paternalistic actions.

I have not taken any explicit stand in regard to my own views on weak or strong paternalistic interventions in general. This is not because I have no opinion on this matter. I have merely refrained from voicing my views on paternalistic intervention because it is not necessary, in the context of my argument, to do so. I tried to demonstrate that, within the theoretical frameworks of the currently debated ethical theories, no ethically convincing case for weak or strong paternalistic interventions can be made that would allow one to argue for such an interference in the particular case of people suffering terminal diseases who are expected to die within a short period of time.[224]

PRACTICAL IMPLICATIONS OF THE ARGUMENTS

In this chapter, we have seen that, when we consider the question of possible paternalistic interventions in the choices of people with AIDS who wish to buy and use unapproved drugs, it is impossible to justify such an intervention given two conditions: (1) the person buying the unapproved drug must have the necessary metaknowledge[225] regarding the risks he or she is going to take, and (2), the person must act autonomously.[226] The practical implications of this decision must also be discussed, however, if this study is to have any real-world relevance. The subsequent paragraphs will discuss those practical problems that I consider important. The premise of the following deliberations will be that my argument has so far succeeded, and that it is illegitimate, on the basis of paternalistic considerations, to prevent persons with terminal diseases from accessing experimental drugs, provided they meet conditions (1) and (2).

Robert Young is one of the few philosophers who has seen the importance of education or economic status for the ability of people with AIDS to make intelligent and responsible decisions that may have far-reaching consequences for the future course of their disease and the rest of their lives. As Young emphasizes regarding external obstacles to autonomy, "Insufficient resources do restrict autonomy. . . . In the extreme case they may even destroy it."[227] Discussions and analyses of bioethicists, who have addressed questions related to the self-determination of people with AIDS or problems related to the participation of such persons in the organization of clinical trials, often ignore these socioeconomic determinants. They discuss ethical problems on the basis of artificial examples that seem to take place in social vacuums. These deliberations ignore that the average person with AIDS is not (anymore) the well-educated, white, middle-class gay man but is often (espccially in the United States) a member of some socioeconomically disadvantaged ethnic minority, such as the Latino IDU.

It seems legitimate that autonomously acting individuals may, at some point, decide that they do not want more information and that they wish to make decisions based on the (maybe shaky and contradictory) information they have already gathered. As long as they have a reasonable understanding of the nature of the risks involved, on a metalevel, it is possible to consider such decisions as autonomous and, to that extent, of a quality that we ought to respect. However, economic, social, and educational inequalities might effectively prevent people from gathering information that other people with the same disease can easily obtain. This can lead to situations in which social factors render patients incompetent. Such a case might involve persons being unable to gather the first-level information they consider necessary to make treatment-related decisions on a metalevel. Under such circumstances, it seems that people are incompetent (perhaps by their own admission) to make an informed decision. I shall discuss possible solutions to this problem in Chapter 3.

In this chapter, I have discussed and rejected the proposition that paternalistic interventions in the lives of autonomous persons with AIDS could be justified. In the next chapter, the primary purpose is to find out whether other, nonpaternalistic arguments might lead us to the conclusion that persons with a terminal disease such as AIDS should be denied access to experimental drugs. These nonpaternalistic discussions are based on an argument which claims that it is desirable for people with terminal diseases to join clinical trials. In order to ensure that they do so, it is necessary to deny them access to experimental drugs, except through participation in clinical trials. The utilitarian justification given for this point of view is that this will help a much greater number of prospective future patients, who will suffer and succumb to the same diseases unless a successful medication or therapy is available.

Chapter 2

Should We Restrict Access to Experimental Drugs to Promote Clinical Trials?

It has been suggested that, without the current participation in clinical trials of patients who suffer from AIDS, future potential patients cannot ever expect to enjoy life-saving medications. To adequately test experimental drugs, so it is argued, we depend on clinical trials with patients who actually suffer from the disease for which we are trying to develop a successful therapy. If patients are given the options of choosing between joining a clinical trial (in which their survival interests do not always have priority over research interests) or simply buying and using the drugs in which they are interested, many potential trial participants would choose this latter option.

There are a number of arguments that attempt to demonstrate why patients should not choose, or should not be given the option of choosing, to opt out of clinical trials. Although such arguments appear in various explications, a common denominator goes along the following lines: "You (current patients) have moral obligations to future generations of potential patients to take part in clinical trials even if these trials bring with them significant disadvantages, some of which might even imply your own death." The reason given for this view is usually utilitarian and argues that these disadvantages, however dramatic they might be in individual cases, are always the lesser of two evils, since they bring about positive benefits for the overwhelming majority of current and prospective patients.

Clearly, it would be problematic to find this argument stated as boldly as I have presented it. Generally, it is not found *expressis verbis,* certainly a direct result of the lack of any democratic consensus in our societies that would legitimate such a strategy. Nevertheless, as I will show, there is little doubt that my assertion of this argument is correct.

Indeed, people with AIDS are aware of this situation, and so are the principal investigators who depend upon these patients' cooperation (which is not necessarily altruistically motivated) in order to run their clinical trials. Ellen Cooper, the former head of the U.S. FDA's Division of Anti-Viral Drug Products, argues implicitly for this position. She concedes that people suffering from a catastrophic illness have a strong disincentive to join a randomized research clinical trial if they have a chance to access the desired experimental agent in some other way. She argues that "a national policy of early widespread availability of unproved experimental agents would slow or even halt the completion of controlled clinical trials through which therapeutic advances are established and then improved on."[1]

The Canadian ethicist John Dixon reformulates this argument as an economic one:

> One of the arguments for sacrificing individual autonomy to those public interests connected with the claims of science focus on the consideration that if we do so we will be able to offer safer, more effective, and hence more economical treatments to future patients.[2]

It is worthwhile mentioning here that Dixon does not subscribe to this economic argument, as I have mentioned in Chapter 1. As his article continues, he presents another possible argument, which holds that society has a right to override individual patient interests in order to do research that is necessary to find a cure or preventative vaccine because "we have an obligation to protect the health of all of society from infection with it, and we have a related obligation to make such progress for the sake of those who may not suffer your disease now, but who will certainly get it and suffer your fate if we cannot discover a cure or more effective treatments."[3] This argument ignores the perception that AIDS is a self-inflicted disease in the majority of cases in the Western world.

It is unclear to me how society can justify expecting individuals to sacrifice vital survival interests in order to help prospective victims of this disease, given the perception that many of these prospective patients could easily have avoided it by, for instance, practicing safe sex.[4] Easily preventable or not, the following practical argument (which I will later support with empirical evidence) remains: Most people with terminal diseases are often unwilling and unable to act altruistically enough to allow scientists to make the progress that is needed at their cost. No matter whether ethical reasoning might lead us to the conclusion that people with terminal diseases should make this sacrifice, if they cannot do it, it is pointless to insist that they should. Ethical reasoning that ignores the social contexts of the issues upon which it pontificates is unsound.

George J. Annas largely concurs with Ellen Cooper's views. He writes:

> Making unproven drugs available is poor policy, [because] if unproven remedies are easily available, it will be impossible to do scientifically valid trials of new drugs. Those suffering from AIDS will be unwilling to participate in randomized clinical trials, and those who are randomized to an arm of the study they do not like will take drugs they "believe in" on the sly, making any valid finding from the study impossible.[5]

Annas unfortunately ignores the fact that persons with life-threatening diseases will not be compliant participants in trials, which implies great setbacks for a number of them. Widespread noncompliance with research protocols has been reported.[6] For instance, patients in the early AZT trials managed to get chemical analyses of the differently colored AZT and placebo capsules. Subsequently, many started sharing this drug with placebo recipients. The trial effectively became unblinded. It is still unclear how many patients used a placebo, how many used the active drug, and at which dosage it was taken. As Martin Delaney reported in the *Journal of Infectious Diseases:*

> AIDS study centers throughout the nation [the United States] tell of widespread concurrent use of other treatments; frequent

cheating, even bribery, to gain entry to studies; mixing of drugs by patients to share and dilute the risk of being on placebo; and rapid dropping out of patients who learn that they are on placebo.[7]

Alvin Novick, a Professor of Biology at Yale University and a member of the FDA's Anti-Viral Committee, supports Delaney's claims in a 1993 article in the *AIDS & Public Policy Journal:*

> Patients dropped out of almost all of the trials presented to us at a rate that appeared to compromise interpretation of the results. . . . Sometimes they (patients and/or primary care physicians) fake entry data, do considerable detective work to identify whether they are on placebo or drug, identify the dosage they received in dose-ranging trials, or otherwise behave actively, by their view of self-interest, rather than as passive subjects.[8]

John Dixon pointed out that:

> more HIV-infected individuals would soon be practicing "preventive treatment" in an effort to save their lives, thus disqualifying themselves as subjects for any controlled and blinded experiments of these new therapeutic options. What is potentially worse, subjects in ongoing trials would be more likely to "cheat"—i.e., to take extra or unapproved drugs without telling the investigators—and thus compromise the experiment. This has already become a bigger problem with subjects in trials of AIDS drugs than medical scientists have ever experienced with any other group.[9]

What Delaney, Novick, and Dixon describe, however, is only the practical consequence of an ethical problem that can be explained as follows: The proponents of the view that patients with terminal diseases should not be given access to experimental therapies do not necessarily hold this view out of paternalistic reasons. Their major motive is the opinion that it is impossible to run clinical trials if such patients are allowed to access experimental drugs. Hence, they believe it is ethically justifiable to block all possible alternative roads to the desired experimental drugs. This leaves patients with terminal diseases with no other choice than to join a clinical trial many would not join if they had the alternative of accessing the desired agent, without the perceived disadvantages inherent to many clinical trials research protocols.

John Arras makes this point quite bluntly. He asks:

> Why should the immediate desires of some for unproven, largely toxic and inefficient drugs outweigh the needs of all HIV-infected persons, present and future, for genuine remedies, especially when the two groups have desires of equal intensity and the latter vastly outnumber the former? *It seems to me that in such a situation a society would be perfectly justified in seeking scientifically validated remedies at the expense of the smaller group's intense desires.*[10]

Obviously Arras' question is of a rhetorical nature. If the problem was as simple as he contends, there would not be any dispute today in the first place. The question is, of course, not whether one set of desires should outweigh another set, but whether we should force persons with terminal diseases to join clinical trials, with significant disadvantages for their own survival interest, through preventing them from accessing experimental drugs.

It is important at this stage of my argument to understand that the point I wish to make is not limited to criticizing nonaccess to experimental therapy. One of the practical implications of this nonaccess policy is the coercion of people with terminal illnesses into clinical trials. Project Inform's Martin Delaney, for instance, takes a firm stand against Arras's and Cooper's view by pointing out that "many patients and their advocates find it morally repugnant to deny potentially life-saving treatment to the masses [in order] to force a few into clinical studies."[11]

This denial of potentially life-saving treatment may either happen as part of the trial design or somewhat accidentally in the course of a given trial. The latter occurs when new treatments are being invented while a trial is being undertaken. If we look at AIDS, for instance, it is important to understand that AIDS is not really a single disease, as is an infection such as hepatitis B. It is a syndrome in which thirty different disease entities are currently identified.[12] Thus, clinical trials are underway not only to eliminate the HIV virus, but also to prevent or successfully treat these opportunistic infections. Many of these opportunistic infections would either not occur or be minor inconveniences in persons with healthy immune systems. They become life-threatening, however, in persons with severely compromised immune systems. Ultimately, it is one, or a combination of these opportunistic infections (or the medications necessary to prevent or treat them) that lead to the death of a person with AIDS.

Imagine now that, at the beginning of a certain trial with a nucleoside analogue, no efficient prophylactics were available for PCP, a form of pneumonia that killed most persons with AIDS. The trial has been set up to run for a period of three years. After nine months, a successful prophylactic for PCP is on the market. If those who were organizing such a trial changed the course of the trial by adding this drug, the trial would lose its significance. It would no longer be clear whether it was the newly developed drug that had led to a prolonging of the average life expectancy of a person with AIDS in that trial or the nucleoside analogue that was being tested. It may also be that an interaction of the agent that prevents PCP with the nucleoside analogue leads to an improved clinical situation of the patient.

Thus, given that the original trial was not designed to answer these questions, the predictive value of the results would be questionable. If the responsible principal investigator did not add this drug to the course of treatment, however, persons in the trial would be knowingly subjected to a course of treatment that was less than ideal. Trials such as this are relatively useless because their results have little relevance for the real-world situation. Patients might understandably switch to successful new therapeutics, regardless of whether they are on a trial that forbids such action. If these new drugs are not intentionally incorporated into a certain trial, it is impossible to say, for instance, whether (and if so, how) they cross-react with the experimental drug under observation in the clinical trial. As we will see later, this was a major problem in the clinical trials that led to the approval of AZT. A certain percentage of persons in the AZT arm had a lower mortality rate compared to the placebo arm. This does not automatically prove that this leads to an overall longer life, as we know now, eight years later. Furthermore, it does not show that it leads to improvements in the quality of life of those persons who are on AZT.[13]

Also, certain populations of people with AIDS (for instance, those in San Francisco) who did not take part in this clinical trial were already, on average, surviving longer than those in either the AZT or the placebo arm of the AZT trial. This was simply due to better clinical care and better management of opportunistic infections, compared to the care taken of those who participated in this particular clinical experiment.[14]

Anthony Fauci, the director of the U.S. National Institute of Allergy and Infectious Diseases (NIAID), makes no secret of his views on this subject. He has reportedly conceded that "the randomized clinical trial routinely asks physicians to sacrifice the interests of their particular patients for the sake of the study."[15] This violates one of the principles suggested by the physician Samuel Hellman, who argued that "individual patients should not be used as a means to achieve even a societally desired end (such as a treatment for AIDS), if in so doing the individual right to medical care is compromised."[16]

Gray and associates largely agree with Fauci's assessment. They argue "that direct benefits generally do not accrue to participants in AIDS research." They acknowledge that "research has benefited humankind tremendously," but they also maintain that the principle of informed consent ensures that those benefits are not obtained at the "expense of the unwilling few." A U.S. Court of Appeals supported this interpretation and ruled, in 1989, that "the needs of the many do not always outweigh the needs of the few or the one."[17] We must ask, however, if consent is really given freely when access to experimental drugs is prohibited in order to obtain a higher rate of participation in research clinical trials.

If the participants in the AZT clinical trial had received optimal patient management instead of, for instance, no prophylactics against PCP, many of them might have lived much longer, with or without AZT. This is one of the reasons why it is ethically problematic to give patients substandard treatments in a clinical trial. Of course, the fewer drugs one tests, the clearer the results, and the more elegant the trial appears to be to peer review. Confusing pictures occur when multiple drugs are involved, and the results will often not be as clear-cut as we wish they were.

These confusing pictures, however, will undoubtedly correlate more closely to patients in the real world. Bruce Mirken, a writer with *AIDS Treatment News,* makes mention, for instance, of a trial testing a recombinant human growth hormone as a treatment for wasting syndrome (a life-threatening opportunistic infection in AIDS). After nearly a year of recruitment, the principal investigator found it difficult to enroll the necessary number of patients, even though many patients were eager to try this hormone. The reason lay in the exclusion criteria the investigators had chosen, among them the use of two widely used drugs, one of which (3TC) remains (at the time of this writing) unapproved. Mirken comments:

> The motivation for such exclusion is usually a desire for "clean" data, uncluttered by the "noise" which might be produced by other drugs—particularly unapproved ones, about which less may be known. But in this case the trial's designers had failed to consider the fact that wasting is generally a manifestation of late-stage disease, and people with wasting were likely to have already gone through all of the approved drugs and moved on to newer ones. They were barring the very patients who most needed and wanted the treatment they were studying.[18]

Ellen Cooper obviously realizes that the design of (many) standard clinical trials might lead persons with life-threatening diseases either not to join them or (as is evidenced later in this chapter) to start cheating because of the obvious disadvantages for a large number of participating persons. She points out that this situation would lead inevitably to a slowing down of the process of advancing therapeutic knowledge. Since she does not advocate the basic restructuring of such trials, it is unlikely that Dr. Cooper's concern is for those patients who actually join them. If it were primarily their individual well-being she had in mind, it would be unclear as to why she would permit the continuation of these sorts of clinical trials, which imply significant disadvantages for many of these patients. Hence, it seems reasonable to assume that her concern is primarily directed toward prospective patients. Such an interpretation explains why she argues that we have to eliminate all alternative possibilities for current patients to access the desired drugs.[19]

Andrew Shorr supports Cooper's argument when he points out that "enhancing the autonomy of current AIDS patients will therefore necessarily involve compromising the autonomy of future people afflicted with AIDS."[20] The claim that we compromise the autonomy of future people afflicted with AIDS by not carrying out research trials today is one I shall question in the next section of this chapter.

RESEARCH CLINICAL TRIALS—ARE THEY DESIGNED TO HELP CURRENT OR FUTURE PATIENTS?

It is perhaps an understatement to say that professional bioethicists suggest that research clinical trials do not take place primarily in order to help the suffering of actual patients. Indeed, they go so far as to claim that these trials take place primarily to prevent potential patients from suffering the same disease. For instance, George Annas notes that the interpretation of clinical trials from the perspective of the principal investigators is very different from the interpretation to which participating patients subscribe. He says, "The investigators see these trials as research designed to provide generalizable knowledge that may help others, while most individuals suffering with AIDS see these trials as therapy designed to benefit them."[21]

As Paul S. Appelbaum and associates argue, this phenomenon of patients misinterpreting the goals of research clinical trials is not limited to patients with AIDS-defining diseases. They contend that what they call the "therapeutic misconception" is "far from uncommon."[22] These authors define a therapeutic misconception as the "denial of the possibility that there may be major disadvantages to participating in clinical research that stem from the nature of the research process itself."[23]

Alvin Novick supports Annas's view through an insightful account of his practical experiences as a member of the Antiviral Drugs Advisory Committee of the FDA. He believes "that a large proportion of patients/subjects volunteered to participate because they wanted access to a treatment that they believed would help them."[24] As I have mentioned already, Andrew Shorr argues that, if we were to give currently living patients the option to access experimental drugs instead of joining placebo-controlled, randomized clinical research trials, these current patient generations would opt out of the trials. This, he believes, harms future patient generations because no reliable drugs could possibly be invented and tested under these circumstances. Hence, it is necessary to ascribe a lower ranking to the survival interests of current patients compared to future patient generations. I do not, of course, suggest that it is argued that each current patient's survival interests are of a lesser importance than those of prospective future patients. It is only argued that, overall, more future patients' interests will have to be weighed against a smaller number of current patient interests.

Hans Jonas has argued, in his famous essay "Philosophical Reflections on Experimenting with Human Subjects," that a society does not have the right to ask patients in current clinical trials to make significant personal sacrifices for the good of society or future patient generations.[25] Even though this position has been criticized recently,[26] it is clear that we cannot expect anybody with a potential or de facto terminal disease to act altruistically in a way that would, in the end, diminish or destroy his or her chances to survive the disease. I agree with Jonas that future generations have the right to a planet, the natural resources of which are intact, and that we have an obligation to give our descendants a chance to live a life which is considered, by them, to be worth living.[27] This means that we have an obligation not to harm them by destroying the environment on which they depend. This does not mean, however, that we have such a significant obligation to do good for them that we must ensure cures for any possible diseases they might suffer.

The view that we (current generations) have obligations toward future generations (of patients) leads to positions such as those of Ellen Cooper and Andrew Shorr. I tend to find the argument persuasive that we have to respect legitimate interests of future generations, insofar as they are affected by our acts today. This cannot possibly mean, however, that we can reasonably be expected to sacrifice ourselves in order to serve the interests of future people, as seems to be implied in the proposals made by Shorr and Cooper. Minimally, it is doubtful that altruistic behavior of this magnitude is practically possible.[28] The actual behavior of persons with AIDS in clinical trials indicates that it is unreasonable to expect them to act in such a way.

Peter Singer argues that we have a moral duty "to prevent something very bad from happening, without thereby sacrificing anything of comparable moral significance."[29] This could be interpreted in our case as supporting the notion that we can expect persons with AIDS to act altruistically in a manner that benefits a much larger number of prospective patients. Tthis seems to be a classical case of a utilitarian counting game (100 patients today versus benefits for hundreds of millions of prospective patients in the future). Indeed, this case has been put forward by John Arras, among others.[30]

Singer is clearly aware, however, that a great deal of the altruistic behavior that would be required in an ideal utilitarian world is beyond the capability of most human beings. Hence, he concludes his discussion of a number of counterarguments against the view he originally advanced, by suggesting a 10 percent donation of our income to support the poor in the world, which is "more than a token donation, yet it is not so high as to be beyond all but saints."[31]

A classic example, with some relevance to this issue, has been expressed by Philippa Foot.[32] She seeks to demonstrate that we have some sort of obligation to actually avoid harming other people and to show that the numbers do indeed count in the real world. She invents a situation in which the driver of a runaway tram has only the choice of steering it from one track onto another. The driver knows that everyone working on those tracks is bound to be killed by the tram if it comes down the track on which they are working. The driver knows also that on one of the tracks only one person is working, while on the other track five men are working. How should the poor tram driver decide? The answer seems clear: if the choice is as described, it would be prudent (all other things being equal) to steer the runaway tram down the track with only one worker.

The conclusion is obviously that (1) we should try to avoid harm to others as much as possible, and (2) if harm will inevitably occur, we should always choose the option that reduces the amount of harm to the possible minimum. However, Foot's example deals with the question of how we should choose if harm will inevitably be inflicted upon others and we are in the situation to decide which of the available options to choose.

Hence, in this case, certain others will always be suffering the consequences of our deeds. The situation clearly becomes more complicated when we change the example a bit. Imagine now that you are one of the workers on the track, and somehow you know that the tram will come along your track if you do not divert it so that it goes down the other route. You are well aware that, if you do divert it, it will probably kill a number of your colleagues, whereas if you do not divert it, only you will be killed. However, having the choice between your own death and that of a number of colleagues, undoubtedly only a member of a rare (and virtually extinct) subspecies of the species *Homo sapiens sapiens* will decide to commit suicide in order to save the lives of colleagues.

This is similar to the situation faced by patients with terminal diseases, when they have to decide whether or not to join placebo-controlled trials. They have to ask themselves: Am I willing to join this trial, at the probable cost of significant disadvantages for my own survival, in order to help other people who would otherwise suffer the same disease in the future? The answer seems to be again, that, if we agree that the numbers matter, then undoubtedly we should join this trial. A Rawlsian, choosing from behind the veil of ignorance about his or her position, would undoubtedly have to make the latter choice too.[33]

However, when we ask ourselves again whether such altruistic behavior can realistically be expected from the average patient, then the answer can only be that we cannot. We cannot seriously expect human beings struggling for their own survival to take their chances with trial designs that neglect the survival interests of many of them. Are we therefore entitled to force someone to join such a trial in order to help others? Is it ethically justifiable to interfere with the individual liberty of terminally ill patients to do what they see as best for themselves and to live their own lives? I do not think that it is possible to create a convincing ethical argument for such a dramatic intervention in the individual lives of persons with terminal diseases such as AIDS.

PHYSICIANS AND PATIENTS

The situation with which we are faced, however, means that a number of persons have to make difficult decisions concerning just who should be involved in clinical trials. Apart from the patient, there is the primary care physician. His or her role is clearly defined as the person whose primary obligation is to take care of the patient's interests. Hence, he or she cannot possibly justify steering the tram down the patient's path or, to continue the analogy, to try to convince the patient to join such a trial. The physician's obligation would actually seem to be to inform the patient about the risks he or she is taking by joining such a trial, wherein the physician might have to inform the patient that joining the trial is not in his or her best interests (though in the interests of future prospective patients). Surprisingly, physicians do not always seem to take the course of action that favors the interests of their patients.

The Swedish philosopher Torbjörn Tännsjö reports that a debate between principal investigators and primary care physicians of persons with AIDS resulted in two groups: a majority who believed that the respective arms of the trial were actually ethically acceptable, and a minority who believed that one of the arms had a much greater (and unacceptable) risk for the participating patients. Tännsjö writes:

> Those who were in the minority, however, found themselves trapped in a genuine dilemma. They felt it was wrong to offer the trial to their patients. But they knew that, if they did not, the trial would still go on with other patients. And now their contribution to it would be that they made it scientifically less valuable. This, in turn, meant that the patients who, in their view of the matter, were sacrificed for the sake of science, were sacrificed for the sake of bad science rather than for good science.[34]

This is a rather odd argument. Primary care physicians essentially have responsibility for their patients and not for any patients who might join the trial because their primary care physicians came to a different evaluation than that of the skeptical physician. Indeed, as Hellman and Hellman ask, "Even though the therapeutic value of the new agent is unproved, if physicians think that it has promise, are they acting in the best interest of their patients in allowing them to be randomly assigned to the control group?"[35]

In any case, if it should become a general rule that primary care physicians advise their patients to join trials that offer substandard treatments to their patients just because they believe that somebody must inevitably partake in the trial, it would seem likely that patients will feel unable to trust such physicians.[36] Perhaps this is the reason why Samuel Hellman argued in another paper that "we lose much more than we gain if we damage the primacy of individual patient care."[37] Therefore, it seems to me that the overall consequences of a general rule that allows physicians to neglect their individual patient's interest for the good of the larger number of patients, or even for any other patient, would lead to a negative outcome, since patients would not be able to trust their physicians any longer.

What might we say about the principal investigator who is not identical with the primary care physician?[38] His or her primary motive is to gather clinical data that have some predictive value for prospective patients with health problems similar to those of the actual patients who have joined the trial. Once he or she has informed the patients of the risks involved and once he or she has received voluntary, and, undoubtedly, informed consent, he or she can indeed steer the tram down any track (given that the workers on either of the tracks were warned of the risk they run when they join the experiment and keep on working on the track). Then, of course, there are primary care physicians, who actively do their own research and who must struggle to fulfill the simultaneous roles of both principal investigator and the patient's doctor.

I have serious doubts that it is practically possible to play both roles without violating the ethical standards of one or both of them. Theoretically, of course, a physician can explain the pros and cons of a trial he or she wishes to undertake with eligible patients. The doctor can explain carefully what the disadvantages are and can try to give a balanced view of the risks and possible benefits the patients might encounter. However, on a more practical level, if the doctor is convinced that the trial is worthwhile, then, of course, he or she will automatically try to convince the patients to join the trial.

Given the influence the doctor has on his or her patients, and the trust they will have invested in him or her, it is likely that the doctor will convince many patients to join a clinical trial he or she has designed, even though under other circumstances, he or she might not advise them to join such a clinical trial. Having said this, it is not surprising that I find it highly problematic in ethical terms for physicians to play the role of the principal investigator and the primary care physician at the same time. On the other hand, who could design a better trial than a primary care physician who is acutely aware of the real health problems his or her patients face?

What about the patients? Kantians would remind us that our acts have to take place out of duty and that the motive for our acts ought to be one that is a universal rule guiding other acts under similar or identical circumstances. Of course, we cannot possibly want no other patients to participate in clinical trials, as Ellen Cooper has warned us, because we realize that this would mean the end of vital research in this area. However, is this really a necessary consequence? This I find doubtful. First of all, it is important to recall that we are discussing a normally lethal disease and not a harmless fungal infection on our feet. It is conceivable and understandable that a person with AIDS wants all current and all future patients with AIDS (or any other illness that is normally terminal) to have, just as he or she has had, the option to decide whether to join a controlled clinical trial or whether to buy and use an experimental drug outside the governmentally approved trials system. Perhaps his or her only intention is to make sure that persons with terminal illnesses have the option to decide for an experimental drug and (for instance) against a badly designed clinical trial.[39]

John Dixon conceded this interpretation in his article on catastrophic rights in terminal disease. He argued that many research clinical trials are sloppily designed and can only be performed because "they didn't have to be sold to a set of medically sophisticated prospective subjects."[40] Dixon believes that if patients are given the choice between access to experimental agents in an unblinded trial versus the option of joining a randomized double-blind trial, those who actually joined the randomized trials would be true volunteers. He predicts that such true volunteers are less likely to break the trial protocol.

To subject principal investigators to this kind of competition might itself suffice to motivate many of them to take greater care in their study design of the participating patient's interests than they have done in the past. This might even lead to a situation where more people than exist currently would be prepared to join such AIDS clinical trials as patients.[41] However, it would also be defensible to hold that, in cases of terminal illnesses where there is an appreciable risk that patients who participate in a clinical trial will suffer irreversible setbacks, the decision about participation should remain entirely in the hands of the patients.[42] Therefore, one could conceivably want as a universally valid rule in a Kantian sense, the decision to participate to remain under the control of the patients, and one could accept, at the very same time, that current and future patients could make practical decisions that one considers wrong or that one would not make oneself under the same circumstances.

The question of whether a primary care physician can justifiably advise any of his or her terminally ill patients to join clinical trials seems indeed warranted. "Bruce," a person with AIDS, has made up his mind. He is skeptical about whether he can trust physicians in clinical trials enough to take the drugs they prescribe.

> I hesitate taking them because I do not know if they do me any good, and I wonder if they may hurt me. . . . I go for awhile and take my medicine and then go for awhile and not take it. . . . I do not participate in drug research that involves a placebo because I want to know what I am taking.[43]

Bruce is not alone in his skepticism in regard to the question of how the researchers tend to his survival interests. Many patients, for instance, those on an e-mail support list for people with AIDS, have echoed his sentiments. One said: "I hate to say it but K . . . we are no more than lab rats to these people. . . . As time passed by [in the study the patient participated in] my [CD4+/8 lymphocyte counts] eventually dropped to almost their original level and I gracefully bowed out of the study . . . thank you Mr. Lab Rat."[44] These patients seem to have good reason to be skeptical. As Novick reported:

> [there have been] many trials that extended over enough time so as to cross clear lines of new standard care for one or more

opportunistic illnesses. . . . There may even have been examples . . . where patients were not advised about wise additions to their care. It was left up to them and their primary care physicians to adjust care that was not specified by the trial: I would presume some had no primary care physician.[45]

Indeed, as I will show later in this chapter, there have even been trials that prescribed knowingly substandard treatments for the participating patients, such as the nonuse of antimicrobial drugs that are necessary to prevent life-threatening opportunistic infections. In addition, I shall point out that trials which offer knowingly substandard treatments to the research subjects violate the 1964 code of the Declaration of Helsinki, as it has been adopted by the World Health Organization. Kenneth J. Rothman and associates noted correctly, that:

the declaration elevates concerns for the health and rights of individual patients in a study over concern for society, for future patients, or for science. "In any medical study," it asserts, "every patient—including those of a control group, if any—should be assured of the best proven diagnostic and therapeutic method."[46]

This general rule, however, is often violated. As a consequence, people are often unwilling to enroll in such a trial, or while participating in it, they are readily willing to start cheating. I will discuss the problems implicated in widespread noncompliance in AIDS clinical trials later in this chapter. Often, study designs exclude the patients that researchers need to recruit by requesting them to accept substandard treatments over the course of the trial. Mirken presents us with just one example of this:

One of the clearest examples of this occurred . . . in what eventually became an important study of recombinant human growth hormone as a treatment for wasting syndrome. After nearly a year of recruitment, the study was enrolling at a painful slow pace. The problem turned out to be the exclusion criteria, which among other things barred anyone taking either d4T or [another drug], both drugs were not yet approved but were available to many patients. . . . After activists successfully lobbied to have the exclusion dropped, the enrollment problem disappeared. The motivation for such exclusions is usually a desire for "clean data," uncluttered by the "noise" which might be produced by other drugs—particularly unapproved ones, about which less may be known.[47]

The consequence is all too obvious, as Dr. Greg Dubs, a person with AIDS and project director of clinical trials, has pointed out: "All you are doing is finding out who can lie well. They never study how many people lie to get into their studies."[48] As I have mentioned earlier, there are significant discrepancies between the interpretations of research clinical trials by AIDS patients and action groups, on the one hand, and researchers on the other.[49]

For instance, in the United States, a country with no universal health care, for many, and especially for poor patients, "the trials offer their [patients'] best chance at good medical care," as Jack Y. Killen, the U.S. National Institutes of Health's director of the National Institute of Allergy and Infectious Diseases AIDS Division, reportedly conceded. Many patients also consider research as some form of treatment. For instance, an advertisement of the Victorian AIDS Council/Gay Men's Health Centre in Australia argued that, "for many people, taking part in a clinical trial may be the only way for them to access promising new treatments."[50] Indeed, Killen realizes that "if they [the patients] could go and get treatment that is currently available we would not be having this discussion."[51]

Subsequently, I shall attempt to demonstrate that in reality we are faced with research clinical AIDS trials that put the patient's survival interests behind research interests. In countries such as the United States, where participation in clinical trials is often the only "health care" poor citizens can obtain, this implies also that the poor take the heaviest burden in terms of the risks involved in any clinical trial.[52] In view of the widespread and serious nature of this problem, I will discuss at some length a research clinical trial that seems to support the claims made by patients about their interests being neglected for the sake of research.

In regard to the trial that will be in question, patient interests were compromised in two respects: (1) regarding the research design that excluded the use of existing prophylactics that were of vital importance for patient survival and (2) in regard to the surrogate markers the researchers chose in order to demonstrate the success or failure of a certain drug over another or over the placebo. In this particular context, I intend to show that the criticism, which was voiced by European researchers against the study's design, was justified. The European critics charged that the surrogate markers used in the Australian study were already discredited. Further, they held that the only reason why these particular surrogate markers were used was to justify the use of a certain drug in an early intervention strategy.

I will now analyze a landmark clinical trial in the history of the United States and international AIDS research. The Fischl/Richman clinical AZT trial[53] (which led to the approval of AZT as an AIDS therapeutic in the United States and de facto to its approval in the rest of the world) is a good example of a clinical trial that I think exemplifies the following two criticisms: it is inherently ethically questionable, and its results are largely useless for a decision concerning whether or not AZT should have been given approval for AIDS therapy.

THE FISCHL/RICHMAN AZT TRIAL

This study enrolled 282 patients with AIDS in twelve participating centers all over the United States, 145 of whom were on AZT,[54] while 137 received a placebo. Of these 137 patients, seventy-five had AIDS, while sixty-two had AIDS Related Complex (ARC).[55] The study was originally scheduled to last twenty-four weeks. The placebo arm was discontinued after a mean of 120 days in both the placebo and the AZT arm. The reason for the premature termination of the study was that nineteen patients in the placebo arm died, compared to only one AZT recipient. The chosen clinical surrogate markers for progress also showed a reduction in the occurrence of opportunistic infections and increases in the numbers of CD4+ cells. The New York City Community Research Initiative on AIDS' medical director, Joseph A. Sonnabend, reported in a critical review of this study:

> The differences in mortality appear dramatic. However, since the efficacy of AZT was measured against death in the group receiving placebo, it is important to know what the causes and

circumstances of these deaths were and that survival in the group receiving AZT can be reasonably attributed to an effect of the drug, that could not have been achieved by any less toxic means.[56]

Sonnabend is here addressing many different points. One is whether it was possible to prevent the deaths in the placebo group by giving them a treatment that is less toxic than the chemotherapeutic AZT. AZT, sold under the brand name Retrovir, leads to severe side effects. Among them are "hematological toxicity including granulocytopenia and severe anemia requiring transfusions," "myopathy, lymphadenopathy, and nausea."[57] Indeed, the *Physicians Desk Reference* mentions even that "significant anemia commonly occurred after four to six weeks of therapy and in many cases required dose adjustments, discontinuation of AZT, and or blood transfusions."[58] Also, given that the AZT group regularly required blood transfusions to counterbalance the hematological toxicity of AZT,[59] it may well have been the case that it was this which boosted the CD4+ levels in the AZT arm of the trial. The principal investigators mention in their own report of this study that "antimicrobial prophylaxis for the prevention of opportunistic infections was not allowed."[60]

One question that obviously needs to be addressed is how a group with an ideal treatment of opportunistic infections and an efficient protocol of prophylaxing these infections would have fared compared to the placebo and the AZT arms in the trial. Such a group, however, did not exist. The study reports that eight out of nineteen deaths in the placebo group were due to pneumonia, a condition for which efficient prophylaxis does exist. It should be noted that at this stage PCP prophylaxis was not FDA approved.

However, the late Michael Callen, co-founder of the New York City People with AIDS Coalition and a patient of Dr. Sonnabend, informed Anthony Fauci, the NIAID director, in an official hearing as early as 1987, that aerosole pentamidine prevents PCP and asked him to issue guidelines for doctors, recommending that they prophylax patients considered to be at risk for AIDS."[61] Fauci refused, as to do so would have been against the FDA regulations. In the meantime, another 16,929 persons throughout the United States died from AIDS-related PCP before the FDA finally approved aerosole pentamidine on June 15, 1989. Nearly 17,000 people with AIDS died unnecessarily because of the FDA regulations. Nussbaum offers further valuable information, in this context, in regard to the research practices of Margaret Fischl. He reports:

> ACT UP also discovered that Margaret Fischl ... was also a PI for an ACTG trial on Bactrim.... Dozens of other community doctors were ... using Bactrim, as well as aerosol pentamidine. Cancer specialists had been using Bactrim and Septra to fight PCP since the seventies. Yet Fischl insisted. Twenty-eight people who received placebo died in her experiment to prove what virtually everyone already knew.[61]

Indeed, since aerosol pentamidine was not FDA approved for AIDS, it was unavailable for the patients in the trial conducted by Margaret Fischl and colleagues. Outside of this trial, many thousands of patients from New York to San Francisco used this prophylactic and found their survival time increased significantly. This survival increase occurred without AZT and was beyond the average survival of patients in both the AZT and the placebo arm. By 1994, AIDS specialists conceded "that PCP prophylaxis and treatment were more important factors in longer survival time than antiretroviral therapy."[62] In this context, it is worth noting that although similar trials are unlikely to take place today in most Western countries, at the time of this writing, they continue in developing countries; for example, the French Agence Nationale de Recherche sur le Sida currently conducts placebo-controlled trials of Bactrim in the Ivory Coast.

Michael Lange, an Associate Professor of Medicine at Columbia University and expert witness of the U.S. FDA's Anti-Infective Drugs Advisory Committee, which had to advise the FDA in regard to the approval of AZT, criticized the design of the trial in this respect. He argued that the trial protocol was designed in such a way that AZT was actually tested against treating pneumonia. Lange was concerned about the trial design because he knew that there were drugs known to work against pneumonia, yet these were not tested against AZT. Lange recalls that "the study was set up in such a way that we were very rigidly controlled about not using any other medication while the AZT was given."[62] Yet drugs such as dapsone and Fansidar were well known as efficient prophylactics against pneumonia.

As I will explain further, this makes the overall results and predictive value of this study doubtful in the eyes of a number of clinical researchers. Investigative U.S. journalist John Lauritsen, whose account of this trial even led to a critical editorial in the medical journal *The Lancet*[64] questioned the validity of the data presented by the researchers by noting that "these mortality data cannot possibly be correct . . . they [are] in conflict with mortality data from other AZT studies."[65]

He gives important information in regard to one of the questions Dr. Sonnabend raised about the causes of the deaths in the study's placebo arm. Lauritsen writes:

> The Phase II mortality data are highly suspect because the deaths were inadequately described. The FDA did not verify causes of death, did not have autopsies performed, and refused to release medical records of patients who died. This is in sharp contrast to procedures used in trials of other drugs (e.g., Ribavirin, where autopsies were obligatory, and a death report form of more than 30 items had to be filled out for each patient who died). There are many indications in FDA documents, released under the Freedom of Information Act, that sicker patients were placed in the placebo group. (L. R. Hauptman, Ph.D., Statistical Review and Evaluation, of NDA 19-655) For example, according to an FDA analyst: "Two patients died very early in the study, one at 10 days and one at 21 days. It is arguable that these patients were sick enough at entry that they should not have been included in the study." Both patients just happened to be in the placebo group.[66]

Documents obtained by John Lauritsen under the U.S. Freedom of Information Act showed that the FDA analyst "who reviewed the toxicology data on AZT recommended that it should not be approved."[67] The supposedly double-blind, placebo-controlled trial became quickly unblinded, as an exposé on AZT broadcast by an American NBC channel in January 1988 revealed:

The test became unblinded almost immediately—everyone knew who was getting what. Patients (in silhouette, voice distorted) told how they could distinguish AZT from placebo by taste. Others admitted taking their medications to chemists for analysis. A chemist described how he had analyzed capsules for participants in the trials. . . . A government memorandum observed that if all patients with protocol violations were dropped, there would not be enough left in the study.[68]

These events could not have happened without, at least in some cases, coming to the principal investigators' attention. It is surprising, therefore, that they still insisted in their publication of the study results that the trial was double-blind and placebo-controlled.[69]

Dr. Fischl and associates put patients' interests second to their research. This strategy has had adverse effects on the participating patients and also on future patients because it has produced unreliable data. Given that AZT was compared with nothing,[70] Sonnabend argued that:

> it is entirely possible that a structured program of patient management might have prolonged the lives of these patients. A structured program of patient management means a formal plan of patient supervision, which includes attempts to diagnose and treat opportunistic infections at the earliest possible time, as well as the provision of general supportive care.[71]

Also, the investigators of this study did not meet ethically important requirements set by the Institutional Review Board regarding the reporting of adverse reactions. An FDA investigator of this study, Patricia Spitzig, notes that the study rules stated, "Any adverse experience by a study subject is to be reported immediately by telephone, followed by a written report." She then reports that "the IRB requirement that all adverse reactions be reported was not met. None of them were reported."[72] Indeed, Sonnabend found significant differences between the causes of deaths reported by the principal investigators to the FDA and those mentioned in the *New England Journal of Medicine* (see Table 2.1).

TABLE 2.1. Causes of Death

Reported to FDA		New England Journal of Medicine	
PCP	4	PCP	8
Toxoplasmosis	2	Toxoplasmosis	2
Cryptococcosis	2	Cryptococcosis	2
MAI	2	MAI	4
CMV	1	CMV	1
Lymphoma	1	Lymphoma	1
"AIDS"	1		
Pneumonia (unspecified)	3	Wasting	1
Pulmonary edema	1		
Suspected MAI, TB, or CMV	2		

Source: Sonnabend, J. Review of AZT Multicenter Trial—Data obtained under the Freedom of Information Act by Project Inform and ACT UP. *AIDS FORUM*, 1989; 1(1):10.

Sonnabend points out in his review of the trial that "it is clear that the causes of deaths were not properly established. The important question of whether any of these deaths might have been preventable, at least within the duration of the study, is therefore more difficult to approach."[73] This, however, seems to be an important question because it turned out to be the case that by September 1986 all patients in the trial were on AZT. Only three months later, there were seven deaths in the group whose participants were originally on AZT. In February 1987, the number of deaths of patients who had been in the placebo arm had risen to eleven.[74] Lauritsen mentions in his review of this trial that all but one of the patients with serious adverse reactions whose problems were concealed were on AZT.[75] Sonnabend quotes from the FDA review:

> Because so many of the AIDS-related signs and symptoms could also be adverse drug experiences, it is difficult to determine whether these events are actually disease-related or drug-

related. It seems that the bias towards reporting them as one or the other (which likely varied among investigators) was altered during the course of the study from a "bias" towards "overreporting" them as possible adverse drug events at the beginning of the study, to "overreporting" them as presumptively disease-associated events later in the study. Adverse experiences were sometimes crossed out months after initially recorded, even though "possibly related to test agent" had been checked off originally by the investigator or his designee.[76]

Sonnabend's problem remains unsolved, as the manufacturer of the drug, the British pharmaceutical company Burroughs-Wellcome[77] concedes in the *Physicians Desk Reference:* "It was often difficult to distinguish adverse events possibly associated with the administration of Retrovir from underlying signs of HIV disease or intercurrent illnesses."[78]

An August 1990 article in *The Washington Post* addresses this problem too, but researchers remained unable to solve it.[79] It is known that AZT is a potent carcinogenic. A review article in the *New England Journal of Medicine* warned as early as 1989:

> In considering early intervention with zidovudine, it is of particular concern that the drug may be carcinogenic or mutagenic; its long-term effects are unknown. It is worth emphasizing again that AIDS itself makes the development of certain cancers more likely, and zidovudine may be associated with a higher incidence of cancers in patients whose immunosurveillance mechanisms are disturbed, simply because it increases their longevity. . . . Lymphomas developed in 3 out of our original 19 Phase I patients between 1 and 2 1/2 years after zidovudine therapy began.[80]

Extrapolation of preliminary data led researchers to calculate that as much as 46.4 percent of patients who take AZT for more than three years, will develop non-Hodgkin's lymphoma. Indeed, a graph in the same study shows at least a correlation between length of AZT treatment and risk of developing lymphoma. The authors concede that "a direct role of therapy itself [in the causation of lymphomas] cannot be totally discounted."[81] As Duesberg pointed out, "The annual lymphoma risk of AZT recipients is about 30 times higher than that of untreated HIV-positive counterparts. It appears that the chronic levels of the AZT at 20-60 μm [500-1500 mg/day] were responsible for the lymphomas."[82]

Others, such as Paul Volberding, head of the AIDS Program at San Francisco Hospital, have argued that "the risk of lymphomas increases the longer a person is infected with HIV" simply because patients with a severely compromised immune system live long enough to succumb to non-Hodgkin's lymphoma.[83] This is supported by a report of the U.S. Center for Disease Control. Even though it conceded that "the pathogenesis of HIV-associated NHL is incompletely understood," it asserts that "the use of AZT is not considered to directly increase the risk for HIV-associated NHL." Furthermore the CDC consider NHL as a problem that occurs "because the life expectancy of HIV-infected patients has been substantially prolonged by more effective treatment of HIV and associated complications, [and] new clinical problems related to prolonged survival of persons with profound immunodeficiency have emerged."[84]

Duesberg, in turn, has argued against this interpretation. He noted that cancers, including malignant lymphomas, are not known to be the consequences of a compromised immune system.[85] A recently published evaluation, however, used the results of a current multistage linkage of NHL among people with AIDS in pre- and post-therapy areas. Cote and Biggar of the U.S. National Cancer Institute concluded that "the relative risk for NHL was marginally higher in the pre-therapy era than when zidovudine was available."[86] The major reason for such uncertainties in our knowledge of the toxicities of the most widely used AIDS drug today is that long-term trials have been prematurely aborted, as the Fischl/Richman trial was, and we simply still do not know with certainty whether AZT is capable of causing this life-threatening type of cancer. The U.S. National Institutes of Health stated, in a recent issue of its *NIH Guide*, that "The median survival of HIV-associated NHL (non-Hodgkin's lymphoma) is less than one year, and is only two months for primary central nervous system lymphoma."[87]

An important implication of the study by Margaret Fischl and associates must be addressed at this stage. A consequence of the argument that current patients can be expected to sacrifice their survival interests for the sake of a larger number of prospective future generations is obviously clinical trial designs that lead to a less than ideal course of treatment for the participating patients. This is problematic insofar as the results of such a trial are only of relevance for patients in the real world, who also receive such a substandard treatment. This is probably not usually the case. Donald I. Abrams, a Professor of Clinical Medicine at the University of California in San Francisco and a member of the U.S. FDA Antiviral Advisory Committee supports this point of view by giving the following information:

> In 1986 the average life span of a patient in San Francisco after PCP diagnosis was 13.2 months, making it a bit premature to truly claim survival prolongation on the basis of this study alone.[88]

Abrams's skepticism in regard to this particular study, which led ultimately to the FDA approval of AZT, is supported by recent findings which concluded that "early FDA evaluations of zidovudine, ... were based [knowingly] on irregularities in patient recruitment and reporting of adverse reactions."[89] It is then all the more irritating that the chief of the National Institutes of Allergies and Infectious Diseases (NIAID), Dr. Anthony Fauci, claims that "the study of zidovudine [AZT] in carefully controlled trials demonstrated conclusively that the drug prolonged survival of persons with AIDS."[90] In an ingenious analysis of this study, Abrams questions whether a decrease in mortality necessarily translates into the prolonged survival that this trial has been given credit for:

> The population of patients with AIDS enrolled in the study had experienced a first episode of *Pneumocystis carinii* pneumonia within the previous 120 days. Only 10% of those enrolled completed the intended 24 week trial; the mean duration of participation was 120 days. Therefore, the AIDS patients enrolled had received their initial diagnoses of PCP only 4-8 months prior to the end of the study. What is clear is that 18 more patients in the AZT group reached the arbitrary landmark of surviving 4-8 months after receiving a PCP diagnosis.[91]

Abrams then points out that at least in San Francisco, at the time when the study was under way, the average PWA who did not use AZT lived several months longer than the duration of the placebo-controlled study.

Indeed, one of the other cornerstones of the Fischl/Richman trial, a rise in CD4+ cells, has been criticized in a number of other studies. One concluded that "levels of CD4+ lymphocytes are an incomplete surrogate marker for progression to AIDS, and the association is especially weak during the first 16 weeks of zidovudine therapy."[92] Subsequently, the largest and longest running AZT study, the so-called Concorde study, confirmed these findings. They studied 1,749 HIV-infected people, who were randomly allocated AZT or a placebo. After three years, the principal investigators summarized their findings:

> Despite the large difference in the amount of zidovudine between the two groups and the fact that the number of clinical endpoints (AIDS and death) in Concorde (347) outnumbers the total of those in all other published trials in symptom-free and early symptomatic infection, there was no statistically significant difference in clinical outcome between the two therapeutic policies.[93]

Furthermore, the authors conclude:

> There was no statistically significant difference in progression of HIV disease: 3-year progression rate to AIDS or death were 18% in both groups, and to ARC, AIDS, or death were 29% (Imm) and 32% (Def) (p=0.18), although there was an indication of an early but transient clinical benefit in favor of Imm in progression to ARC, AIDS, or death. However, there was a clear difference in changes in CD4+ cell count over time in the two groups. . . . Thus *persistent differences in CD4+ cell count do not necessarily imply long-term differences in clinical outcome.*[94]

I will explain briefly the harmful implications of the idea that CD4+ counts are an important surrogate marker for progression of disease. It is another consequence of the outcome of the early AZT trial by Margaret Fischl and associates. For instance, in reporting about a very recent clinical trial with the protein interleukin-2, a Reuters news agency dispatch said: "Tests on 26 volunteers infected with HIV show that interleukin-2 can raise the number of vital CD4+ white blood cells by an average of at least 50%. . . . [However], there is no evidence to show whether the treatment may help infected patients to live longer, better lives."[95] The very fact that a surrogate marker such as CD4+ can be raised steadily is being hailed as evidence that the immune system is stimulated, even though there is no evidence that this will improve the length of survival of people with AIDS or that it will even improve their quality of life. In fact, the principal investigators mention in their own summary of the study results that "[In 15 patients with CD4+s ≤ 200] therapy was associated with increased viral activation, few immunological improvements, and substantial toxic effects."[96]

Abrams mentions another trial:

> This dissociation of endpoints appears to have occurred once more in the large randomized European/Australian study code named Alpha, which compared high and low doses of ddI in 1,930 patients intolerant to AZT. A small, but highly significant, CD4+ benefit was seen in the high-dose group, yet survival and disease progression rates were equivalent in the two groups.[97]

Indeed, in the same article Abrams quotes the authors of this particular study as saying, "The lack of difference cannot reasonably be attributed to chance; this study is powerful statistically because of the trial's size and the large number of events."[97]

The problems in regard to attempts to find useful surrogate markers (see next paragraph) are quite significant. The debate about CD4+–rises as a meaningful marker for clinical progress is only another example of such problems. The latest fashionable surrogate marker in AIDS is the viral load in people with AIDS. Common sense would suggest (just as in the case of CD4+/CD8 counts) that there is a direct connection between viral load and disease progression. In fact, many physicians have, at the time of this writing, already started to initiate or switch therapies on the basis of viral load. This, however, may again be premature. As Gregg Gonsalves of the New York City-based Treatment Action Group has pointed out, "There is little evidence to suggest that small or moderate, treatment induced changes in viral load will correlate with a clinical benefit of antiretroviral therapy. . . . A small or moderate drop in viral load for a few months may not be predictive of longer disease-free time. It is still to be proven."[98] Gonsalves warns that the drugs used to achieve these decreases in viral load are toxic, and the balance between their beneficial and deleterious effects remains unclear:

> In many other diseases, "beneficial" changes in common-sense "surrogates" due to therapy, like viral load do not necessarily correlate with better clinical outcome. In cancer some chemotherapeutic regimens can shrink the tumor, but kill the patients on it faster than those who are not; in heart disease, some drugs can lower cholesterol or reduce arrythmias but provide no survival benefit or actually reduce survival compared to placebo.[98]

What are the classical criteria for a meaningful clinical endpoint in a research clinical trial? The four criteria most widely accepted in the medical profession are:

1. First and foremost the primary endpoint should be relevant and easy to interpret—an obvious example is survival. If a treatment extends survival and does not adversely influence morbidity or quality of life, the decision by clinicians and patients of which treatment to use is easy. If the endpoint is an event short of survival, then it should be highly correlated with survival;
2. the endpoint should be clinically apparent and easy to diagnose. It is particularly important to ensure that disease surveillance is identical for treatment groups, otherwise bias may result;
3. the endpoint should be sensitive to treatment differences—if there are true and meaningful differences between two treatments, the primary outcome measure should capture them. This characteristic of sensitivity is particularly important in

studies comparing two or more active agents for which treatment differences are not likely to be large; and lastly
4. the endpoint should be measurable within a reasonably short time frame after randomization, if possible, so that the relative efficacy of new treatments compared to standard therapies can be established rapidly. The desire for quick answers concerning effectiveness against specific clinical conditions has to be balanced against the possibility that the answer derived has little relevance. Differences in the long-term effects of treatment on survival, for example lead to different treatment recommendations than those based on intermediate clinical conditions.[99]

As the authors of this review paper correctly note, "this is, of course, one of the concerns in using surrogate markers, such as CD4+ lymphocyte counts, as the primary outcome variable."[99]

It is surprising that a trial such as the Fischl/Richman study has led to the approval of AZT through the U.S. Food and Drug Administration. Indeed, the shortcomings of this trial have been articulated by a number of commentators. This is, perhaps why, even today, the *Physicians Desk Reference* states under the heading Retrovir, "Zidovudine has been studied in controlled trials in significant numbers of asymptomatic and symptomatic HIV-infected patients, but only for limited periods of time. Therefore, the full safety and efficacy profile of zidovudine has not been defined, particularly in regard to prolonged use."[100] In medical journals, however, the manufacturer of the drug has advertised that the indications on which AZT may be prescribed are based on the results of a "randomized, double-blind, placebo-controlled trial conducted at 12 medical centers in the United States."[101]

The harmful implications of this particular trial have been significant. A number of lawsuits in the United States and Europe, against both the manufacturer and those governmental institutions responsible for the drug's approval, are pending. Some of these cases relate to other AZT trials that have been stopped prematurely on grounds similar to those which were accepted by the FDA from the principal investigators of the Fischl/Richman trial. For example, another trial, which was supposed to analyze the impact of AZT in early therapy (before the onset of any AIDS-defining diseases), was terminated prematurely when NIAID director Dr. Fauci was convinced of the beneficial effect of the drug. A lawyer acting for the widow of a British AIDS victim claimed, according to a newspaper report, that medical records indicate that Dr. Fauci's announcement led her husband's doctors to start him on AZT.[102] The results of much longer-running clinical trials have been published in subsequent years, and not one trial, which ran over its originally planned period of time, has confirmed the results of the trial that led to the FDA approval of AZT in early therapy. In fact, a very recent review article of major AZT studies—the results of which have been published between 1993 and 1994— reported that not one found survival benefits.[103]

Obviously, we are faced here with another ethical problem, as Paul Volberding has pointed out. Volberding, a Professor of Medicine at the University of California at San Francisco and the Chief of the AIDS Program at San Francisco General Hospital, asks, "Is it fair to require volunteers to continue to take a placebo or the inferior drug in order to demonstrate differences in overall death rates?"[104] Indeed, can we seriously expect volunteers in such a trial to continue taking a drug that seems to be inferior to another agent just because we as a society have a keen interest in understanding the long-term effects on survival of a certain agent? Isn't it understandable, and perhaps even reasonable, for a person with a terminal illness to accept indications of some sort of benefit, even though the information is less than complete and the long-term outcome is unknown? I am not competent to judge about whether there are technical alternatives. For the sake of the argument, I will assume that there is no alternative than to go ahead with the trial at the cost of quality of life and ultimately the survival of a number of patients. The only intelligible answer to this problem that I can imagine is to continue the trial with patients who are truly volunteers (as distinct from desperate patients who have had no alternative other than to join the trial).[105] Under the current legislation that prevents access to experimental drugs for patients with terminal illnesses, it seems to me that there is no way to justify ethically the request that they obey the protocol because they are not volunteers in the first place.[106]

The patients' lawyers argue (supported by some medical professionals) that the severe toxicity of AZT has, in fact, contributed to the outbreak of AIDS in persons who were only HIV positive before they started using this drug. Legal action has also been taken on behalf of Susan Threakall, a woman who believes that her husband's death was caused by AZT. "[She] says her husband was only given AZT because of unjustified claims for the drug by Wellcome [the drug's manufacturer] and the United States National Institute of Allergy and Infectious Diseases."[107]

ETHICAL ANALYSIS OF THE FISCHL/RICHMAN AZT TRIAL

This study, so obviously methodologically flawed in many vital respects, has determined the FDA decision to approve AZT in AIDS therapy.[108] The ethical problems to be discussed are related to questions of design and the reporting and interpreting of results by those who conducted the study. The major reason given by the principal investigators for the early termination of this study was that the ratio of deaths in the placebo versus the AZT arm was so significant that it would have been ethically irresponsible to continue the protocol. This view has been accepted by the Anti-Infective Advisory Committee to the FDA, as this body's chairman, Itzhak Brook conceded. He reported in the *Journal of the American Medical Association* that the trial in question was aborted prematurely, after preliminary data indicated that AZT increases the short-term survival of the participating patients. His committee was asked by the FDA to consider the merits of Wellcome's application for approval of AZT as an AIDS drug even though not all relevant "clinical data, exact dosages, long-term follow-up and possible long-term toxicity were available."[109] Brook defended the FDA's position by pointing out that the decision to continue the evaluation process in the absence of some of the relevant data constituted an exceptional step and that it was a direct response "to the desperate need of the thousands of patients with AIDS for whom no known cure is available."[109]

This argument for accepting insufficient data to release a drug with admittedly serious toxic side effects claims to have the needs of thousands of patients in mind for whom no successful standard therapy is available. Thus, Brook presents us with an ethical claim. He is basically apologizing for approving a drug that under normal (read non-AIDS) circumstances would not have been approved by the FDA on the basis of the evidence then available. Indeed, he said so explicitly when he opened the debate on AZT for the members of the advisory committee: "Because of the rushed nature in which medicine and industry are trying to cope with it [AIDS], we are having to discuss a drug that may not have been studied as adequately or as thoroughly as the FDA is used to having drugs studied."[110] AZT received approval because patients were in desperate need. Obviously, patients do not need a drug that is neither efficacious nor safe. To the extent that the efficacy and safety of AZT were doubtful, the patients' need for AZT approval was also doubtful. The only reason why this drug was approved, it seems, is because patients put tremendous pressure upon the FDA to do so. Bruce Nussbaum reports that gay AIDS treatment activists pressured members of Congress to approve a drug to treat their disease. At the same time, the Reagan administration was eager to show that the United States, not France, was in the forefront of the international efforts to find a treatment for AIDS and/or a preventive vaccine. Given these circumstances, it seemed politically opportune to cut the trial short.

It is important to understand at this moment, however, that patients are only interested in seeing a drug such as AZT approved because this was the only condition under which they were guaranteed access. They would not have minded waiting for certain results, e.g., of the Concorde clinical trial (which failed to support the claims made in the Fischl/Richman trial), had they been permitted to access experimental drugs such as AZT at their own risk, while waiting for the results of a clinical trial (or indeed to access drugs that are not clinically investigated at all). This interpretation of mine is supported by a recent editorial in *The Lancet*. It reported that "AIDS activists in the USA have lately urged the Food and Drug Administration (FDA) to delay approval of new anti-HIV agents" because neither efficacy had been established nor toxicity levels had been sufficiently determined. The editorial continues, claiming that this is in "extraordinary contrast to the situation in 1990, when the FDA, under great public pressure, licensed the early use of zidovudine for symptomless HIV infection."[111] The editorial fails to understand that patients were only interested in getting a final approval for a drug of unknown efficacy and toxicity for particular reasons. Namely, this was only a consequence of the existing drug approval regulations. For instance, Derek Hodel, a Washington, DC-based lobbyist for the U.S. AIDS organization Gay Men's Health Crisis, wrote a commentary in regard to the already mentioned attempts by the newly elected conservative majority in the U.S. Congress to curtail the regulative powers of the FDA. He wrote:

> AIDS activists will occupy an unusual position in this debate, as much of the charged rhetoric now employed by the conservative right substantially echoes the arguments crafted by activists themselves during past well publicized debates with the agency [the FDA] concerning the approval of AIDS drugs. Many activists are understandably ambivalent about rushing to the defense of the government agency that had been the focus of their ire only a few years ago. But most AIDS organizations also agree that their intention was never to debilitate the agency, and like many more traditional consumer groups, they are now worried that the so-called reform proposals [suggested by Republican members of Congress] would do just that—and what is worse, might actually diminish the number of effective drugs developed in the future.[112]

Patients whose autonomous decision making was already severely curtailed due to these drug approval laws saw no other alternative than to push for the release of a drug that was not proven to be either efficacious or safe by pre-AIDS scientific standards.[113] So, it seems that the policy of prohibiting access to experimental therapies by patients with terminal diseases has led to clinical trials such as the Fischl/Richman trial and to the approval of drugs that are neither proven to be safe nor efficient. Indeed, the aforementioned *The Lancet* editorial supports this conclusion by pointing out that the precedent set by this particular trial has led to other trials being terminated prematurely too. The editors wrote:

> The decision (to approve AZT for use in HIV-infected symptomless individuals) was largely based on randomized clinical trials that were terminated early—e.g., the ACTG019 trial, which in 1989 was stopped on the grounds of significant response over placebo for patients with CD4+ cell counts

under 500/µL, even though mean follow up was only 1 year. The larger European Concorde trial continued its planned randomization and follow-up despite these results, and the preliminary report showed no clear difference in survival or disease progression, with a mean follow-up of 3 years. Thus disillusionment began to set in.[114]

By this time, however, hundreds of thousands of people throughout the Western world with an HIV-positive test result were on AZT monotherapy. This was undoubtedly a result of this inappropriate legislation.

Such an outcome has undoubtedly negative implications for future patient generations because they are offered approved drugs that, in the end, do not offer the safety and efficacy they might reasonably expect from a medication that has been approved by the U.S. drug approval authority. I have mentioned already the case of Susan Threakall suing the U.S. NIAID and the manufacturer Burroughs-Wellcome for having the drug released on the basis of insufficient evidence of its efficacy and, in particular, knowledge about its toxicity. Subsequent studies, which accumulated many more patient-years, failed to confirm the results of the Fischl/Richman trial, and more important, they failed to confirm the basis on which the FDA decision that led to the approval of AZT in AIDS therapy rested.[115]

An ensuing debate in the medical journal *The Lancet* concluded with statements such as this: "The lack of effect on mortality of early use of zidovudine seen in these seven trials . . . is in stark contrast to the pronounced benefit observed in AZT" in the trial conducted by Margaret Fischl and colleagues. Furthermore, the authors point out that "trials lasting less than 18 months appear to show some benefit of immediate zidovudine while those lasting more than two years suggest a possible adverse effect."[116] Indeed, a study published in March 1995, in *Annals of Internal Medicine*, suggests that, for a subgroup of AIDS patients using AZT, the drug might lead to an overall shortening of their life expectancy. It seems that those who develop resistance to AZT are three times more likely to die within a year's time than those who either have not taken the drug or whose viruses remain susceptible to AZT.[117]

It can, unfortunately, even be shown that trials such as the Fischl/Richman study have had far-reaching implications for other drugs under consideration. Carlton Hogan, a member of the University of Minnesota's Division of Biostatistics, reports:

> Perhaps more disturbing . . . is the failure of our methods: A state of the art [SOTA] conference at the National Institutes of Health reviewed all the available data on nucleoside analogue reverse transcriptase inhibitors [NARTIs]. For the majority of situations the panel came up with, they could not recommend any specific indication as to when, how, or even if NARTIs should be used, despite a billion [U.S.] dollars spent researching them.[118]

The American ethicist Gregory E. Pence is one of the very few bioethicists who have not fallen for the positive hype that initially surrounded the use of AZT in AIDS therapy. In his most recent book, he notes that "in 1994, the Concorde study was followed by more discouraging results about AZT."[119] At this point, it is important to emphasize that the reason for the accelerated approval of drugs with basically no efficacy data, is a consequence of an inherently unethical situation. People with AIDS are forced to join clinical trials in order to access experimental drugs they wish to use. Given that these designs do not even allow all participating patients access to the desired drug, and that this is the only legal road to these drugs, widespread cheating is an all-too-human consequence.

The result is, as I have pointed out earlier, harmful to future patients with AIDS who have to rely on essentially untested drugs. This unfortunate reliance on questionable clinical trial results is a direct consequence of preventing persons with terminal diseases from accessing experimental drugs, which in turn forces them to join such a trial in order to access the drug they believe might save their lives and which also puts pressure on government authorities to speed approval of some form of treatment for a widespread terminal disease. AIDS activist groups seem largely to agree with my interpretation. A coalition of five French groups of AIDS patients and activists has argued:

Some people consider those who leave trials in such large numbers [the group reports earlier in their statement that in two recent large clinical trials between 34 and 42 percent of all patients dropped out of the trial for reasons not anticipated in the protocol] irresponsible in a certain manner since by their individual decision they endanger the research whose results will serve some, if not all, HIV-positive persons. But self-sacrifice has its limits; rational collective logic is impossible for each individual to consider. In addition, with HIV-infection, medical follow-up is done more and more frequently in the context of one or several trials. So, for HIV-positive people, clinical research and health care are often blurred. Sometimes the rigidity of trials combines awkwardly with the evolution of the practices already in place. Thus, substandard treatments in certain trials are to be genuinely feared. Such people, more and better informed all the time, are no longer ready to accept this phenomenon. This is the tension at the heart of the current clinical research crisis.[120]

The Fischl/Richman trials are also ethically problematic on a number of other levels. At the time they were conducted, AZT seemed to be the last hope for many thousands of persons with AIDS all over the United States. Still, only a very small number of them had the chance to join this trial, due to a limited number of available trial places. A major ethical problem that occurs is the question of the voluntariness of the informed consent given by the participating patient subjects, considering that they had no other way to access this experimental drug because all alternative (legal) roads to AZT (or other experimental drugs) were blocked by U.S. legislation. Of those who finally did make it into the trial, nearly 50 percent received a placebo. Undoubtedly, however, receiving a placebo was not what they had hoped for in joining the trial. Hence, as aforementioned, an unknown, but seemingly significant, number of patients began to share drugs without informing the principal investigators. Thus, the trial became neither blind nor placebo controlled. As a scientific study it was rendered questionable.

Other studies have had to struggle with another phenomenon, namely patient withdrawal. As the French AIDS organizations mentioned earlier pointed out, "The clinical Phase III trials of antiretrovirals seem completely maladapted to the patients' living situation, testified by the tremendous number of premature withdrawals seen today in most of the current large trials of this sort."[121] Both the consequences of cheating and patient withdrawal might have been avoided had patient autonomy been respected, as I proposed earlier in this book. In such a case, patients would have joined the study voluntarily, and the chances that they would have complied with the protocol requirements would have been much greater. After all, what incentive would someone have to deceive a principal investigator in a placebo-controlled, double-blind clinical trial if he or she could access the desired drug without committing fraud?

Indeed, the very choice of CD4+-counts as a surrogate marker (as opposed to ultimate survival) has had harmful implications. Donald I. Abrams argued persuasively that, "we really must face the critical issue [of prolonged survival] and not pooh-pooh survival in favor of modest transient increases in CD4+-counts."[122] A review article by Thomas R. Fleming, "Surrogate Markers in AIDS and Cancer Trials," confirms this. Fleming points out that, before one can accept a surrogate marker as a replacement endpoint, one needs to be sure that "the marker fully captures the effect of treatment on the true endpoint." His review of sixteen of the major AZT trials led him to conclude, "It is very apparent that one cannot simply consider establishment of statistically significant treatment effects on CD4+ cell counts to be a valid surrogate for either of the two clinical endpoints [development of AIDS or death]."[123] Fleming stresses that "the relationship of CD4+ effects and survival is even more unsatisfactory [than that of CD4+ effects and development of AIDS]."[123] Indeed, another study by DeGruttola and associates concluded that the effect of AZT on CD4+ levels predicts very little, if anything at all, about the drug's effect on survival.[124]

One of the final ethical points that needs to be discussed in regard to the Fischl/Richman trial is that it required patients to accept substandard treatments. An example of this is that patients did not receive "antimicrobial prophylaxis for the prevention of opportunistic infections."[125] It was well known, even at that time, that persons with AIDS do not die of AIDS per se but of opportunistic infections. Because of this, many primary care physicians throughout the United States started to prevent and treat opportunistic infections to the best of their ability. Indeed, it has been argued (for instance, by Professor Abrams, whom I have quoted above) that the increase in the average survival of a person with AIDS after the development of the first disease symptoms is largely due to this treatment strategy, rather than (early) treatment with nucleoside analogues such as AZT. When the AZT trial conducted by Margaret Fischl and associates started, this was already common knowledge. Therefore, patients received a less than ideal course of treatment in this trial. Sonnabend asks:

> Is it ethically acceptable to withhold an intervention, or a series of interventions—in this case good patient management—in order to test a particular treatment for the underlying diseases in a trial that measures efficacy by its ability *alone* to reduce the occurrence of an opportunistic infection or death? For example, should pneumocystis pneumonia prophylaxis be withheld in order to test the efficacy of an antiretroviral drug?[126]

I have mentioned earlier in this book the problem of so-called "therapeutic misconception."[127] When this occurs, patients believe (even though the informed consent form they signed tells them the opposite) that the primary function of the research clinical trial in which they take part is to help them survive. At this stage, we do not know how many of the patients in this particular trial were actually under such a misconception. Still, it seems fairly obvious that few, if any, persons with AIDS would voluntarily and knowingly take their chances in a trial design that is effectively refusing available treatments for the prophylaxis or therapy of potentially life-threatening microbial opportunistic infections. One possible means of explaining why it seems that people did just this invokes the therapeutic misconception. In addition, it might well have been the case that patients believed that the possible benefits of AZT, in a manner akin to a magic bullet, would outweigh the disadvantages of not receiving prophylactics. If such a hope in AZT did exist, however, it would surely imply that the incentive for placebo patients to deceitfully access the experimental drug should have been considered formidable. Moreover, such a pretrial belief in the properties of

AZT would almost certainly entail some form of the therapeutic misconception, whereby patients come to believe that they can access a magic cure by entering a trial, while forgetting that the emphasis of a trial is to test a drug whose benefits are unknown. Furthermore, regardless of the existence of the therapeutic misconception, it remains ethically problematic to motivate people to join trials that offer substandard treatments, when such motivation depends on their inability to access the experimental agent at their own risk.

Finally, the nature of this research protocol also means that its results are not very useful for future patient generations. After all, real-world patients might understandably go to their physicians in order to receive antimicrobial prophylactics. These prophylactics, in turn, may well cross-react with AZT (or any other drug). Thus, the clinical trial conducted by Fischl and associates is not giving any information in regard to whether or not AZT would prove efficient under real-world circumstances. Therefore, it remains all the more doubtful in ethical terms as to whether the disadvantages the patients were supposed to accept (nonprophylaxis against microbial agents) are justifiable.

The ethically relevant implications of the Fischl/Richman trial can be summarized as follows:

- Unreliable data are used in order to propagate a certain drug.
- Patients are unable to make informed decisions between the course of treatment proposed in this study and alternative courses of therapy.
- Patients in the study were subjected to substandard treatment.
- Even if the data of this study had been reliable, it is doubtful that these particular data presented the ethically relevant information. A major recent study about the quality of life people with HIV enjoy, when they are on AZT, concluded that the "reduction in the quality of life due to severe side-effects of [AZT] therapy equals the increase in the quantity of life associated with a delay in the progression of AIDS."[128]

The last point is supported in real-world decisions made by people with AIDS or those caring for children with AIDS. Associated Press carried a story that was broadly supportive of the conclusions put forward by Lenderking and colleagues. Associated Press reports that the adoptive parents of a three-and-a-half-year-old, HIV-infected child made the decision to take her off AZT in the fall of 1992. The result was that the child "appears healthy and no longer experiences the upset stomach, appetite loss, and severe leg cramps associated with the antiviral drug."[129] The mother argued quite adamantly that she would rather have her daughter live another two months with a good quality of life than subject her to six bad months under the chemotherapeutic AZT.

In Australia, a gay man with AIDS who decided (against the advice of his physician) to discontinue taking AZT remembers, "I was on AZT for 6 months. Although the dose was only 600 mg a day, I suffered horrendous migraines, loss of appetite, and blue flashes, before kicking it and going alternative. Then I felt like I wanted to die—now I feel 100%."[130] These two examples are representative of a general trend as far as AZT is concerned. A financial report of the Reuters news agency recently stated that the U.S. sales of AZT in December 1993 were 25 percent lower than in the same month of 1992.[131]

I have provided an analysis of the Fischl/Richman trials and presented information about a number of other problematic trials. This discussion, in conjunction with the ethical commentary, has led me to conclude that:

1. Patients with AIDS are expected to join clinical trials that offer substandard treatments.
2. Such ethically problematic trials lead to questionable results and may even harm future patient generations, whereas a trial that takes more account of the interests of the (basically involuntarily) participating patient subjects would lead to more useful results.

The questions that really needed to be answered in regard to this trial and perhaps a number of others are, in my opinion:

1. At what stage (if any) of disease progression is AZT prolonging life (assuming that it is not a cure for AIDS)?
2. At what stage (if any) of disease progression is AZT prolonging disease-free life?

3. How does AZT affect the quality of life as it might be given at the stages determined in 1 and 2?
4. Is it possible to gather information in ethically acceptable clinical trials that would allow patients to make informed, autonomous decisions in regard to AZT?[132,133]

Hardly any of the trials thus far conducted have helped the formulation of an adequate answer to even one of these questions. One trial, the results of which have been published recently, concluded that AZT given to asymptomatic HIV-positive people with more than 400 CD4+ cells per cubic millimeter results in benefits lasting more than two and one half years.[134] However, this trial was handicapped by all the known problems of large numbers of patients (apparently many more in the AZT group) withdrawing from the trial.

A critical article by Phyllida Brown was published subsequently in the *New Scientist*. Brown discusses the implications of this particular study for prescription rules for AZT, and she makes a few critical remarks about how Burroughs-Wellcome, the manufacturer of AZT, has seized on this study in order to improve sales and to dispute the results of the Concorde clinical trial. (It might be emphasized that the Concorde trial proves to be a notable exception to those that have addressed none of the aforementioned questions.) An editorial comment in the same issue of the *New Scientist* notes that:

> the acid test of a drug trial is whether it changes doctors' practice. Concorde has done so: a panel of independent scientists in the U.S. has just tightened the recommendations for who should receive zidovudine, effectively reducing the number of patients who will take it in future. Not surprisingly, Wellcome has repeatedly disputed Concorde. The same panel also discussed the Australian trial, but did not consider it important enough to change its recommendations for treatment. Wellcome may be excited by the trial, but few others are. The company should know better than to hype something of so little importance.[135]

Wellcome has been blamed for "mischief making and of being disingenuous in its portrayal of the findings."[136] It is interesting to note that David Cooper, the principal investigator of this study, which was fully sponsored by Burroughs-Wellcome, is not known to have made any attempt to correct the misleading use of his study by the company that financed his research. Ian Weller, the chairman of the Concorde clinical trial pointed out that "the results of Concorde are completely consistent with [Cooper's study] for CD4+ cells, but what Concorde shows is that CD4+ changes are not a very good marker of disease progression."[137] This conclusion has been subsequently supported by biostatisticians such as Susan Ellenberg. Ellenberg pointed out that "the data from the Concorde trial tell us that treatment effects on CD4+ counts do not accurately predict the extent of clinical benefit and that this marker therefore cannot be used reliably to discriminate among active treatments and identify those that will produce the greatest clinical benefit."[138] A study by Schlumpberger and colleagues reported that CD8 counts have a higher predictive value for survival than CD4+ counts. They conclude, "for every 100-cell decline in the enrollment CD8 count, the risk of death increased by 16%, *independent of other enrollment variables, including CD4+ counts and p24 antigen status.*"[139]

Recently, the journal *Genetica* published a critical statistical evaluation of Cooper's assertion that AZT provides a beneficial effect in terms of progression to AIDS for HIV-positive people who are asymptomatic and have CD4+ counts higher than $\geq 500 \text{mm}^3$. The author concludes:

> Statistical analyses of the published data of the EAGCS by CD4+ cell count range were performed to test the conclusion of the EAGCS [Cooper's group] that AZT therapy is effective in delaying the onset of clinical symptoms of AIDS when given to those with baseline CD4+ cell counts $\geq 500 \text{ mm}^3$. The results of the analysis do not indicate that there is a statistically significant association between AZT therapy and progression to any of the four independent endpoints of the EAGCS for either CD4+ cell count ranges, 500-749 mm^3. Nor does the analysis indicate a statistically significant association between AZT therapy and progression to any of the four independent endpoints for the combined group, with CD4+ count range $\geq$ 500 mm^3. . . . These analyses indicate there are no benefits

from AZT therapy in terms of progression to AIDS for those who are asymptomatic and have CD4+ cell counts that are ≥ 500 mm^3.[140]

Professor Weller was reported in the London *Sunday Times* to have mentioned another problem with the use of surrogate markers that are not directly correlated with clinical benefits or better survival. He argued that "research of this [Cooper's] sort was aimed more at raising sales than knowledge." The *Sunday Times* article continues: "Weller whose previous academic post was sponsored by the Wellcome Trust, parent of the drug-manufacturer—said that if Concorde had not been supervised by government agencies he believed it would have been stopped at a point advantageous to AZT. This has occurred with all other tests of the drug."[141]

Subsequently, a number of correspondents expressed their skepticism in regard to this study in a letter to the editor of the *New England Journal of Medicine*. They concluded that "the assertion that zidovudine has a beneficial effect that lasts for more than 2 1/2 years in these patients is not justified on the basis of the results presented,"[142] and "in terms of the date on progression to AIDS or ARC, the group given zidovudine and the group given placebo differ by only four patients, in whom *Pneumocystic carinii* pneumonia developed. No individual case data are presented on prophylaxis against *P. Carinii* pneumonia."[143]

COERCIVE OFFERS, OR YOU CANNOT HAVE IT BOTH WAYS

It has been argued repeatedly that it is not the function of drug approval authorities such as the U.S. FDA to actually develop drugs and offer them to the population. The function of such institutions is primarily to prove the safety and efficacy of drugs that pharmaceutical companies wish to sell. This is undoubtedly prudent. After having shown in the first chapter that it is impossible to justify preventing persons with terminal diseases such as AIDS from accessing experimental agents on the basis of paternalistic reasons, I have argued in this chapter that it is also impossible to justify preventing these patients from accessing experimental drugs on the basis of the argument that this causes actual harm to others.

Protagonists or defenders of the current status quo have subsequently presented other proposals, which I will criticize in this section on the basis that they constitute ethically unacceptable coercive offers. One of these proposals holds that there is no such a thing as a moral right of people with AIDS to join particular clinical trials. This view rejects the request of patient advocates that all people with AIDS (or other terminal diseases, for that matter) should be permitted to use experimental drugs within the framework of clinical trials. The idea here is that patients should not be given access to experimental drugs outside clinical trials, but this view recognizes, at the same time, that all patients should be offered the option to join a clinical trial.

It is, however, practically impossible to admit every patient who wishes to join a certain specific protocol (because not enough hospital beds are available or because a number of exclusion criteria have to be met).[144] It is probably even impossible to admit most patients who request a place in clinical trials. Nussbaum observes that only a few hundred people were enrolled in clinical trials in a state with hundreds of thousands of HIV-infected people.[145] In addition to limited availability of places, there is also a recruitment problem. Often enough, willing patients are available en masse to join ethically designed clinical trials, but those who organize the trials are simply out of touch with real-world patients and fail to recruit them in an efficient manner. ACT UP reportedly perceived the reality of clinical trial participation and recruitment in this way: "The NIAID wasn't telling any people with AIDS anything about its own trials.[146] Thousands of people with AIDS were trying to access experimental drugs by means of participation in clinical trials, yet the NIAID didn't tell them how to do it. Nussbaum reports that the trial enrollment was left in the hands of principal investigators who had to recruit the necessary number of participating patients. "Yet it was clear from the accrual numbers in the July NIAID report that the PIs were doing a miserable job of getting those patients."[146]

Defenders of the current situation have reacted to this criticism along these lines: "We try our best, of course, to give as many patients as possible access to our trials, but our means to do so are limited and so is, after all, the number of patients that can participate in our trials. Given that no patient has a moral right to join any such trial, our offer objectively improves this patient's baseline situation. He or she had no chance of survival before, and the offer to join our trial has de facto improved the patient's baseline situation. Nobody, however, can force us to make this generous offer to any patient who might be willing to accept that same offer."

In the words of John Arras:

> Although much obviously remains to be done in *improving* the baseline situation of potential subjects with regards to the improved availability of experimental drugs, their present lack of access is probably best described . . . as "unfortunate but not unjust."[147]

Arras borrows this phrase from H. Tristam Engelhardt. Engelhardt argued that situations, which are comparable to some sort of natural lottery in the sense that they treat all participants equally well or badly, are morally neutral. Indeed, he points out that "providing inadequate health care is not *prima facie* unfair or unjust."[148] This example, however, does not really fit our problem. The situation we are finding ourselves in is not a natural situation; rather, it is human-made. And this human-made situation is clearly disadvantaging persons from poor economic backgrounds who cannot afford to disobey the law.[149] Also, as we have seen earlier, minority groups, such as IDUs, are still discriminated against by the medical research establishment within the overall population of persons with AIDS.[150] Hence, we cannot even assume that, within this state-run lottery, all those who would like to take their chances are being allocated the same chance of winning.

This is one reason why I think Arras's point of view appears to be convincing only at first glance. Once we have a closer look at the real situation of people with AIDS, we come to realize quickly that those who make these coercive offers are part of a medical research system that has brought the patient into this seemingly hopeless situation in the first place.[151] The clinical research system and the drug approval legislation effectively force the patient to join a clinical trial that is not primarily interested in the participating patient's survival.

The current legislation prevents a terminally ill patient from buying and using experimental agents that this patient, or his or her primary care physician, considers worth trying. This brings the patient into the hopeless situation of either accepting the friendly offer of the principal investigator or dying. This amounts to a de facto denial of a reasonable option to choose in a situation that is of vital importance for this patient's future life. It is somewhat surprising that ethicists such as Macklin and Friedland continue to argue under such circumstances that "those who agree to be subjects in a controlled trial in which they may be randomized to a placebo group do so, presumably without coercion and after being fully informed."[152] Indeed, even German physicians such as Wolfgang Wagner, the editor of a handbook on pharmaceutical ethics and the corporate medical director of a major pharmaceutical company, have argued in a similar vein that the decision of people to join a clinical trial is always an autonomous one. Furthermore, they argue that it follows that the patients therefore accept that the physician's duty to benefit them will be limited for the sake of gathering knowledge in a given clinical trial.[153]

A state that truly allowed its citizens with AIDS to make their own choices would have given them the following minimum options to choose from:

1. joining a placebo-controlled clinical trial;
2. joining an intention-to-treat clinical trial;
3. buying and using the desired experimental drug; or
4. doing nothing and waiting to see what happens.

Should the state decide to limit the patient's options to the possibilities of joining one or another kind of clinical trial (1 and 2) by making option 3 illegal, then we have clearly crossed the threshold to a coercive offer. A real choice between different options can no longer take place because it is all too obvious to both the principal investigator and the patient that option 4 will almost certainly lead to the death of the person with AIDS. The state coerces, almost by default, all those patients with an interest in their own survival to join a clinical trial that many of them would not join if all other roads to experimental drugs had not been blocked by the legislators. This subtle force also leads to patients giving consent to research designs they would presumably not accept if they were left with other options.

It is indeed questionable whether it is ethical to speak of patients having given their *voluntary* informed consent to participation in a given clinical trial under these circumstances. For instance, in the case of a hijacked airliner, in which the hijacker allows persons to choose between being killed first or watching the other passengers being killed one after another, we would not claim that the passengers have chosen voluntarily to join this process. Similarly, which person who suffers from a terminal illness would give voluntary consent to nontreatment (a placebo)?[154] No matter how significant the risk related to the use of an experimental agent in a clinical trial, it appears more rational to accept this risk than to choose a placebo.[155]

This is why I cannot agree with John Arras when he writes, "The question is not whether potential research subjects have 'no choice,' but rather whether AIDS researchers or 'society' have an independent obligation to provide HIV-infected persons with better alternatives than those in current protocols."[156] Clearly, this statement gives answers to more than one question. We might indeed ask whether society has an obligation to provide persons with AIDS with better alternatives. If one disagrees with Arras's negative answer, one could imagine a wide range of offers society could make to persons with terminal diseases. An example might involve the development of programs allowing for wider access to experimental drugs under the control of principal investigators or primary care physicians, wherein efficacy and toxicity data were still collected.

On the other hand, one might concur with Arras's conclusion, and hold that the state has no such obligation. Accepting this argument, however, does not entail that the state therefore has the right to prevent people from taking their lives into their own hands. It is one thing to say that the state has no obligation to give experimental drugs without sufficient scientific control (as provided in clinical trials) to persons with terminal diseases such as AIDS. It is an entirely different matter to claim that the state has the moral right to actively prevent such persons from buying and using such drugs at their own risk and expense.

Even if we were to accept the argument that neither the state nor the pharmaceutical industry have an obligation to provide terminally ill people with chances to survive (perhaps through the offer to join a clinical trial), then it seems only consistent to reject the proposal that the state has the moral right to prevent such people from taking their fate into their own hands. The current political climate implicitly suggests that one of society's obligations to its citizens is to take part in drug research, either actively as an institution undertaking research and/or as an institution that controls the outcome of such research. Even in situations approximating this one, however, it does not follow that the state also has the right to expect extreme altruistic behavior from its citizens. This is another reason why Arras misses the crux of the problem when he argues:

> No one is obligated to make promises, but once made they must be kept. Likewise no one has the duty to become a research subject in the first place, but by entering a protocol, subjects enter into a moral relationship with researchers by promising . . . to abide by certain restrictions in return for the benefit of participation. Dropping out of an ethically designed study merely because one has to be randomized to placebo, pooling drugs, or taking unapproved remedies all amount to violations of this promise.[157]

My reasons for rejecting Arras's point of view are as follows:

1. Promises made in situations of acute emergency need not always be kept by those who make them. They certainly need not be kept, when these promises have been given to those who are responsible for this emergency situation in the first place.
2. Patients with AIDS, as well as a respectable number of primary care physicians, would deny that many of the current AIDS clinical trials are ethically designed. This implies that we are faced with a situation where such people would, at the very least, reject Arras's conclusions because their interpretation of what constitutes an ethical design differs from that of Arras.

PLACEBOS AND OTHER DRUGS

I shall briefly discuss one final argument in this chapter. It does not apply to situations where my overall argument has been accepted, i.e., that it is unethical to prevent patients with terminal illnesses from using experimental agents outside the context of clinical trials. Rather, in addressing this final argument, I am concerned with multiple-arm trials that offer patients either (1) the chance of receiving an experimental agent or a placebo or (2) the chance of receiving different experimental agents (which might include a placebo). As mentioned previously, it has been argued in both ethical and legal literature that a trial is always ethical when it assigns the volunteers the same chances and risks regardless of the arm to which they will be randomized. The late Benjamin Freedman has described this situation, where there is a genuine uncertainty among the researchers, as a "clinical equipoise."[158] An example of such a trial might occur when the chances for the patient that he or she will fare better with the drug are the same as those for the patient who is receiving a placebo. An alternate case might involve the chances for a given patient in arm I, who receives agent A, being just as good/bad as they are for a patient in arm II, who receives agent B.

I would like to make a general point regarding the argument that invokes clinical equipoise as a criterion for judging the ethicality of a trial. In essence, this assumption involves the claim that principal investigators have no indication as to how a certain experimental agent destined for phase III testing in a clinical trial will work in actual patients. I find this claim questionable in the majority of cases. Obviously, the major reason for testing a new experimental agent is that it "promises more effectiveness," as Samuel and Deborah Hellman pointed out.[159] Clearly, there are principal investigators who would insist that they have no opinion at all as to whether participating in the placebo arm or the arm with the experimental agent is the better option. However, I think the Hellmans have made a convincing point, when they argue, "If the physician has no opinion about whether the new treatment is acceptable then random assignment is ethically acceptable, but such lack of enthusiasm for the new treatment does not augur well for either the patient or the study."[160]

As a matter of fact, many preclinical tests have been undertaken before an experimental agent is finally tested in a clinical trial with human subjects. These tests must already have given a clear indication in favor of the drug. Had results been otherwise, the drug would not have been considered for a clinical trial in the first place. Hence, in the case of placebo versus experimental agent, the principal investigator's assumption has to be that the chances are higher that the experimental agent works than that the placebo works. Such an assumption in favor of the drug, however vague, must question the likelihood that clinical equipoise is achieved often, if at all. Moreover, to the extent to which it is achieved, it bodes well for neither the trial participants nor the study and, as such, provides reason against considering such a trial to be an ethical enterprise.

Perhaps a similar argument can even be made in the case of principal investigators who test multiple drugs (with no placebo controls) in a number of trial arms. It may well be that the investigators have no absolute knowledge of the outcome. After all, they would not run the trial otherwise. However, it is very likely that they have, based on their knowledge of the preclinical data, an assumption that would allow them to rank the experimental drugs in order of the likelihood that each might work.

SUMMARY

To summarize the major conclusions of this chapter, one can have at least two views in regard to whether the state has the duty to actively do research in order to develop courses of treatment against diseases that threaten its citizens. If one holds the view that it is not the obligation of the state to do so, it is reasonable to argue that the state has no right to prevent its citizens from trying to find a cure on their own and to use unapproved drugs or unapproved courses of treatment. This argument is particularly plausible in light of my objections to weak and strong paternalistic interventions in cases of terminal illnesses, as presented in the first chapter of this book. However, even if one holds the more common view that to do such research falls within the range of duties the state has toward its members, then there is still no justification for holding that the state has the right to recruit trial participants via coercive offers of the kind discussed previously.

It also seems to be the case that, within this already unethical clinical trials system, it is even more reprehensible for the state (and principal investigators) to offer knowingly substandard treatments to people with terminal illnesses in the course of clinical trials. This is so because it is not possible to provide a persuasive argument which would allow us to assume that the extreme altruistic behavior required on the patient's side is ethically justifiable. Hence, the actual harm inflicted upon patients who are required to respect protocols (and obey protocol rules) that offer such substandard treatments is unacceptably high. Furthermore, it appears that trials requiring such substandard treatments (or, indeed, admission criteria that systematically exclude large subgroups of patients such as IDUs) actually harm future patient generations. This occurs in at least two respects:

1. If patients did indeed stick to the trial protocol, then we would only know how a certain experimental agent works in patients who receive substandard treatments. At first glance, this might seem a tautology, but it describes one important implication of such trials: they can only give us information about how a certain course of treatment works in people who might receive similar substandard treatments in the real world. Normally, however, we would not find such patients often (if at all) outside of clinical experiments. Those patients who actually receive substandard treatment commonly do so because their primary care physicians are incompetent or because they simply cannot afford better care. Thus, future patient generations (who do not have the misfortune to be receiving similarly substandard treatment) might be harmed, insofar as trial results are at best irrelevant and at worst misleading for the sort of effects they can expect from a given drug.
2. Patients in such trials are known to cheat in large numbers. This occurrence leads, for instance, to the unblinding of these trials, the sharing of drugs with placebo recipients, and patients dropping out of trials in large numbers. These consequences lead to results that are methodologically questionable.

Finally, it seems that one of the arguments in favor of placebo-controlled trials (that is, that the principal investigators have no well-grounded opinion as to which of the trial arms is most likely to provide benefits) is not convincing. In actual fact, principal investigators must have such an opinion because, otherwise, their reasoning for testing a certain drug would be unacceptably weak.

Ultimately, trials need to focus on survival or on surrogate markers that are proven to be linked to survival or, at the very least, to clinical benefits. We need trials that mirror real-world patients, not artificial and unrepresentative patient groups such as those I have mentioned in regard to a number of trials in the preceding chapter. As Hogan pointed out in this context, "The prevailing trial model, with restraints on eligibility, concomitant medications and clinical management can only hope to accrue and retain people very unlike most people with HIV or AIDS. . . . The most urgently needed information today is that which will guide real-world treatment choices."[161] A number of alternatives to the current gold standard of clinical research, the randomized clinical trials, have been proposed.[162] We urgently need clinical trials that answer the right questions and not trials which show us yet again that it is possible to increase the CD4+ levels with a certain agent. Susan Ellenberg has explicated the reason for this very clearly. She wrote:

> Utilizing surrogate endpoints [such as CD4+s] . . . requires the assumption not only that biological activity [e.g., increase or decrease in the number of CD4+ cells] results in clinical benefit, but that the degree of clinical benefit can be precisely predicted from the measure of biological activity. The legitimacy of such an assumption for the use of CD4+ counts in AIDS trials is far from established.[163]

Still, there is more that needs to be said in regard to research strategies. I believe that it is practically possible to transform a situation in which patients choose to use experimental agents into a viable research strategy.[164] The advantage is clearly that the information which we gather is real in the sense that it is not the product of a trial protocol which does not reflect real-world living conditions of people with AIDS. I suggest that, even though people with AIDS should be given access to experimental agents of their choice, they should only be able to do so after consultation with a physician. This is consistent with the arguments that I have presented so far. The physician's function is not to prevent people with AIDS from accessing these drugs. Rather, it is to inform them about the pros and cons of the drugs in which the patient is interested and to gather information about how the patients fare with the experimental agent. This includes information about positive and negative effects, as well as data about the dosage, frequency of intake, results of experimental desensitization protocols, etc. In other words, far from meaning the end of clinical research, access to experimental agents can actually be an important complementary means of gathering information. These data, in turn, should be fed into data banks to ensure that the experiences of hundreds of thousands of people with AIDS can and will be shared with their fellow patients and with clinical researchers.

Chapter 3

Costs and Other Practical Problems

I have demonstrated that it is impossible in virtually all cases to justify preventing terminally ill people from buying and using experimental agents on the basis of paternalistic motives. Furthermore, I have argued that respect for autonomy is of significant importance because dispositional autonomy is a necessary instrument in our individual attempts to live our own lives. Toward the end of Chapter 1, I discussed some of the specific handicaps people with AIDS face and made a few practical proposals regarding how to improve the level of individual competence and decision making in these patients.

I analyzed the question of whether the current laws (as represented in most Western societies), which prevent people with terminal diseases from accessing experimental agents, can be ethically justified on grounds other than those which are paternalistically motivated. Although the latter are usually only present implicitly, explicitly they accept that the participation of patients in clinical trials is essential to achieve the knowledge required to provide safe and efficient drugs for potential future patient generations. Hence, according to these arguments, it is necessary to prevent patients from using experimental drugs because this would prevent them from joining clinical trials in which their participation is urgently needed. I rejected this reasoning in the second chapter of the book. In addition, I argued that this strategy has, in the past, led to the substandard treatments of people with terminal illnesses and, as such, that it is ethically unjustifiable. Finally, I pointed out that such clinical trials ultimately lead to results that are largely useless under real-world circumstances.

Two practical questions remain. First, who should cover the costs that would inevitably occur if the wider community accepts my argument? Second, who should be in charge of the aforementioned information facilities (databases, etc.) to be set up for people with terminal illnesses?

In order to answer the first question, it might be useful to invoke a general point to emphasize my view that there is a broad international consensus about health care provision to citizens through governmental agencies. The international community of states, in the form of the United Nations, has explicitly stated the right of each human being not only to health, but also (as might be of more practical importance) to health care in its classical declarations and covenants regarding human rights. For instance, the *Universal Declaration of Human Rights*, as it has been accepted by the general assembly of the United Nations, declares, "Everyone has a right to a standard of living adequate for the health and well-being of himself and his family, including . . . medical care."[1] The *Universal Declaration of Human Rights* is little more than a declaration of the UN member states that they intend to live up to the goals set out therein. The declaration is in no way legally binding. It is only a moral obligation accepted by all states that are members of this international body. In practical terms, the specifications set out in the International Covenant on Economic, Social and Cultural Rights of 1966 are probably more important. The signatories of this covenant agree to implement the content of this paper in their national legislations. The International Covenant unequivocally states:

> The States Parties to the present Covenant recognize the right of everyone to the enjoyment of the *highest attainable standards of physical and mental health*. The steps to be taken by the State Parties to the present Covenant to achieve the full realization of this right include those necessary for: . . . c) The prevention, treatment and control of epidemic, endemic, occupational and other diseases; d) The creation of conditions which would assure to all medical services and *medical attention in the event of sickness*.[2]

The World Health Organization (WHO) considers the individual's interest in sufficient health care to be a fundamental human right. The WHO, following the International Covenant's formula, writes, "(the) enjoyment of the highest attainable standards of health is one of the fundamental rights of every human being."[3] Unfortunately, this formula is ambivalent and allows for various contradictory interpretations, as Heta and Mati Häyry have shown.[4] However, a right to be healthy becomes somewhat meaningless if a right to access to adequate health care does not accompany it. Most countries that have a functioning welfare system acknowledge this. As Robert Goodin correctly pointed out:

> A state that did not even try to relieve distress when it was clearly within its power to do so could not credibly claim to constitute a welfare state. . . . For a state to qualify as a welfare state, it must attempt to relieve distress in such a way that the agents of the state are bound by a legal duty to provide those in distress with certain resources that they need.[5]

Indeed, Desmond S. King and Jeremy Waldron concluded their essay "Citizenship, Social Citizenship and the Defence of the Welfare Provision," by arguing that "social rights of citizenship are a necessary condition for genuine and meaningful consent to social and political arrangements." It is reasonable to assume that a society that does not provide such a basic welfare system to its citizens alienates "those who happen to be poor or untalented." These citizens might claim, "this society does not treat us as members, as citizens, or as people who belong here, for it is operating a system which could not possibly have commanded our consent." As King and Waldron correctly note, "if an argument like this can be made, then there is a reason for having welfare rights in a society whether it has a tradition of such institutions or not."[6]

A moral right to health is to be interpreted as a justified claim of a given individual not to be hindered by the government, or responsible health care institutions, in his or her pursuit of health. Furthermore, the state or government is required to support such striving for healthiness within its means. In the context of this book, it is impossible to analyze in adequate depth the question of whether quasi-natural moral rights exist or whether they are a purely rhetorical means in political debates.[7] I consider the arguments of those who dispute the actual existence of such moral rights quite convincing.[8] Nevertheless, whatever one's view regarding this question, one thing is beyond doubt between the different ideological factions in this area: The claiming of X as being a moral right usually has a primarily political function. In democratic societies, there seems to be a broad consensus that what is morally good should be implemented in the law, unless the alleged immorality is one of a type that Feinberg has called a "harmless immorality." This term refers to an act that is generally considered immoral but which is harmless in that it does not hurt parties other than those directly involved. However, Feinberg is at pains to point out that the "offense principle cannot justify the prohibition of 'offensive' conduct even where it does not offend, without undergoing metamorphosis into the unpalatable principle of legal moralism." Therefore, Feinberg argues that "the offense principle cannot be used as a life-raft to save the ship-wrecked legal moralist."[9]

It can be established that there is more than a mere chronological nexus between the societal acceptance of certain extralegal norms and their subsequent implementation into legal norms. Indeed, it may well be that this connection is, in many cases, a causal one, as Reinhold Zippelius has suggested.[10] The rhetoric of moral rights essentially has the function of clearing the way for the implementation of the alleged moral right as law. If there is a legal right established, citizens can usually both refer to and rely on it when they demand certain services from the state. They have, if necessary, the legal system of a given democratic country on their side, as opposed to a government that refuses to abide by the law. Debates are ongoing over how far-reaching the moral and legal obligation of a given country's government should be in terms of providing health care for its citizens. Undoubtedly the answers will be different for each society, given that they depend strongly upon how the relationship between the state or society and its citizens is defined.[11]

The motives for the interest of states in taking a degree of care of the health of its members (citizens) are multilayered. Certainly there are a number of reasons that are anything but altruistic, such as the understanding that a society can operate successfully and prosperously only if the vast majority of its members are healthy and able to fulfill their daily duties in their respective workplaces. The richer a given society is, the more workers it can afford to lose for a limited period of time, without threatening the existence of the society as a whole. Strong paternalistic governments, such as that of Singapore, have presumably acted out of such motivations in running extensive public campaigns against smoking. The Republic of Singapore has finally banned it in virtually all public locations (including the public transport system, restaurants, and discotheques) and the workplace.

All democratic Western-style societies accept that the lifestyle decisions of particular citizens might generate higher costs for all members of their societies. This status quo is generally based firmly on ethical reasons.[12] Of some importance for understanding this view is that the United Nations civil and social rights covenants require governmental respect for individual autonomy and the right to self-determination. Hence, the statement, "All people have the right to self-determination. By virtue of that right they freely determine their political status and freely pursue their economic, social and cultural development." The International Covenant contends "that these rights derive from the inherent dignity of the human person."[13] Health is a necessary (though obviously insufficient) condition for the realization of individual rights and obligations by autonomously acting beings. As we have already seen, philosophical attempts to justify our respect for individual self-determinations are manifold. Authors from divergent philosophical traditions, among whom are authors as diverse as Immanuel Kant,[14] John Stuart Mill,[15] or more recently Gerald Dworkin,[16] have provided us with a variety of more or less disparate theories, all of which stress the importance of respect for individual autonomy.

Another motive can certainly be derived from thought, which has encouraged human beings to overcome the anarchy of the lawless state of nature. Jean-Jacques Rousseau describes such changes in his *Social Contract:*

> The passage from the state of nature to the civil state produces a very remarkable change in man, by substituting justice for instinct in his conduct and giving his actions the morality they had formerly lacked. Then only when the voice of duty takes the place of physical impulses and right of appetite, does man, who so far had considered only himself, find that he is forced to act on different principles, and to consult his reason before listening to his inclinations. Although, in this state, he deprives himself of some advantages, he gains in return others.[17]

States acknowledge that certain minimal justice principles concerning the distribution of goods among its citizens generate a generally better result (in regard to the satisfaction of the average citizen's interests) than other systems. Widely accepted principles of justice and other ethical principles substitute the so-called law of the nature. Perhaps because of these reasons, we find that a system of government-organized public health care exists in most of the Western industrialized nations.

In all countries of the European Union, as well as in Australia and Canada, universal health care is guaranteed, even to those members of these societies who cannot afford to join a private health care fund. In a landmark decision in 1969, the American Medical Association declared, "It is the basic right of every citizen to have available to him adequate medical care."[18] At the time of this writing, however, citizens of the United States of America have neither a guaranteed constitutional nor a general statutory legal right to health care.[19] The consequences of this are particularly devastating for the poor who suffer from AIDS or other terminal illnesses. Some forty million Americans are without any health coverage, while another twenty million are underinsured by any reasonable standards of adequacy.[20]

Gregory Pence gives this account of the current situation in the United States:

> Does the average independent American physician ration care? Yes and no. If a patient has a generous medical plan, the physician may be able to order anything at all, and the plan will cover it. If the patient has a managed care plan, the physician may be limited by what the plan will cover and thus may be unable to order the "ideal" treatment. If a patient is underinsured or uninsured, the physician may order or recommend a certain treatment—such as a prescription drug—but the patient may simply decide that it is unaffordable and forgo it.[21]

The situation in some Southeast Asian countries that are experiencing rapid economic development (such as Singapore or Malaysia) is actually better than that in the United States and is more comparable to that of countries such as the United Kingdom, Australia, or Germany.[22] It appears that democratic countries, as well as outspoken autocratic-paternalistic societies, have decided that one central responsibility of the state consists in guaranteeing and promoting the health of each of its citizens. This is independent of other considerations or dispositions, such as whether these citizens are rich or poor or whether they are members of a certain ethnic group.

Nevertheless, despite good ethical arguments and arguments derived from international human rights covenants, it is obvious that there is no international consensus in regard to whether states have a moral obligation to supply health cover to their citizens. For the subsequent discussion, I will therefore distinguish roughly between two different types of societies:

1. those societies with universal (state guaranteed) health cover for all of their members;
2. those societies that require individuals to take care of their health cover on their own.

The first question was: who should cover the inevitably occurring costs of access to experimental drugs for people with terminal illnesses? The second question that needs to be addressed is certainly: who should be in charge of the information facilities to be set up for people with AIDS or other terminal illnesses? In this context, I will briefly address other important practical issues such as how the experiences of people who use experimental drugs can still be used in order to further our knowledge and, in a certain way, to continue drug research.

ACCESS TO EXPERIMENTAL DRUGS IN SOCIETIES WITH UNIVERSAL HEALTH CARE

Certain societies, such as the German or the Australian, by and large, guarantee universal health care to their citizens.[23] Within such societies, finding an ethically justifiable answer to the question of whether people with terminal illnesses should be allowed access to experimental drugs is not the only problem that has to be addressed. Thus, critics of the idea that access to experimental drugs should be allowed have argued that, even if we accepted the arguments that I have presented in the preceding chapters, a number of interrelated questions still remain open.

Such questions might be: Who should pay if terminally ill people use experimental drugs with unknown effect on their health? Should society pay for drugs that might not work at all or that may even be harmful? Can the obligatory insurers, such as those in Germany, who cover everybody who is not privately insured, seriously be expected to pay for experimental agents of which little or nothing is known at all? What if these drugs do more harm than good? What if these people actually end up in the hospital earlier than they would have if they used standard treatments? These are difficult practical issues that need to be addressed, especially if one finds my arguments persuasive in favor of allowing access to experimental agents.

As Robert Young has illustrated in his book *Personal Autonomy*, it is clear that external constraints, such as the insufficient financial resources of a given patient, can severely limit or even destroy this individual's (dispositional) autonomy. Young wrote:

> Suppose R to have substantial financial resources at his disposal and S to have much less. R is able to choose a life plan, which would involve him spending large sums of money on, for instance, travel entertainment and other desirable trappings of life. S is unable to choose. Whether or not R chooses thus to spend his money, he (unlike S) has the option. R is to that extent better able to live autonomously. Insufficient resources do restrict autonomy. In the extreme case (as of abject poverty) they may even destroy it.[24]

It should be clear that in a society that gives people with terminal illnesses such as AIDS the permission to use experimental drugs, the same consequences are not experienced by each patient. This is precisely because we do not live in an egalitarian society where all patients have the same capabilities and starting points. Some patients may be highly educated, thus better able to make informed choices than others. Other patients might be wealthy and thus able to use their considerable resources in order to gather information, while others (poor IDUs, for example) could not afford to do so. Wealthy people would also be able to buy expensive agents that remain beyond the reach of poor people, who depend on a public health care system that does not subsidize the acquisition of experimental drugs.

In societies with universal health care, I argue that patients can reasonably expect the health care system to pay part of the bill for the experimental agents they wish to use and to cover the costs in cases where the experiment the terminally ill patient has undertaken goes wrong. This is not because these situations are special. A terminal illness is not, in and of itself, sufficient to support the claim that a state with universal health cover should foot the bill for experimental agents. However, even in cases of terminal illnesses, there are treatments in use that are actually paid for by the state-subsidized health care system. This means that the government is already willing to pay for agents that are proven not to save the lives of patients. Hence, I think it can be argued that in order to increase patient autonomy in cases of terminal illness, such societies can be expected to pay for those experimental agents that patients might prefer over the drugs for which the state-supported system normally reimburses. The logic for this claim is that, since there exists no successful therapy, it is intelligible to argue that patients should be allowed to make the difficult choice between a standard treatment that is not working and an experimental agent that might well work. Furthermore, it is arguable that they are entitled to be respected, and even supported, in their decision.

Such support might be given in the form of patient-vouchers, equivalent to the amount of money that is being spent per patient treated in the conventional way.[25] A system of vouchers is used, e.g., in the United States, where no universal health care exists, to allow poor people to buy the food necessary for their survival. The owners of the vouchers might be permitted to spend them with a health care practitioner of their liking, including practitioners who might not have government approval, such as those using alternative therapies.[26]

Arguments for such a support scheme can be derived from the principle of equal respect for interests of people in similar circumstances.[27] As Peter Singer points out:

> We weigh up interests, considered simply as interests and not as my interests, or the interests of Australians, or of people of European descent. This provides us with a basic principle of equality: the principle of equal consideration of interests. The essence of the principle of equal consideration of interests is that we give equal weight in our moral deliberations to the like interests of all those affected by our actions.[28]

There is no fundamental difference between the interests of a terminally ill person who chooses a standard form of treatment that is known not to save his or her life, and a similarly ill person who chooses alternative treatments whose efficacy is unknown. In fact, in terms of survival, one might even argue that it is more reasonable to take a risk with an unknown drug, instead of taking a drug that is known not to work. Therefore, the principle of equal respect for equal interests seems to require that we actually reimburse the costs incurred by patients who choose to use experimental drugs.

So far I have only argued for an affirmative answer to the first question that I posed, which was whether the governmentally organized health care system can reasonably be expected to pay the bill for a reimbursement of costs that are incurred in the acquisition of nonapproved agents. What about additional costs that might occur when an experimental agent not only does not work, but actually causes harm? What if a person ends up in a hospital earlier than he or she would have, had the person decided to take the approved drug? My answer to this may seem inhuman at a first glance, but it is a direct consequence of the argument presented in the chapter on paternalism. I argue that no real additional costs are actually incurred. What happens is that people with a terminal illness injure themselves significantly and that they deteriorate faster than they would have under normal circumstances. No real additional costs are actually incurred simply because, if people deteriorate faster, they will die faster than people on standard treatments. Hence, if the cost argument were to be taken seriously in this context, it would ultimately be in the best interest of such a society to allow people to die faster in order to save money.

It is important to note that this argument does not apply to therapies that have been proven to work. I am only arguing for a situation in which patients suffer from a terminal illness. We have no reason to expect society to pay for experimental agents in cases of diseases that are actually curable and for which successful therapies are available.

I will now discuss and refute a few other possible counterarguments to my proposal. Most of these arguments try to demonstrate that, when we take Singer's argument for the equal consideration of equal interests seriously, we have, as Singer himself writes, to examine the interests of all those affected by our actions. The first counterargument must be seen against this background.

One could argue that allowing access to experimental therapies for all people with terminal illnesses, who are capable of acting autonomously, prevents societies from running successful clinical trials designed to test certain drugs. I have discussed this argument at some length in the preceding chapter. In the context of the question of who should pay for experimental agents outside clinical trial protocols, it could be argued that, if we allow this access, we violate the survival interests of a much larger number of prospective patients for whom no treatments will ever become available because we cannot run carefully designed clinical trials. To that extent, the principle of equal consideration of the interests of all those affected by our actions might lead us to try to motivate people to join clinical trials. Clearly, if society pays for their experimental agents, current patients have one less reason to join a clinical trial. Hence, the argument goes, society should not reimburse costs that occur when these people buy and use experimental agents. The costs that should not be subsidized are at least twofold:

1. those that occur directly in the acquisition of the agent;
2. those that occur when the agent does not work and the patients need medical attention that may be expensive.

This argument might concede the points that I have made in the preceding chapters. Its proponents may not attempt to prevent people with terminal illnesses from using experimental agents through legal means. They will argue, however, that because this behavior is not in the best interest of a much larger number of people who will come into a similar situation in the foreseeable future, nobody can expect the taxpayers of this society with universal health care to sponsor the use of experimental agents. Thus, although the citizens of this society might be willing to respect the personal autonomous decisions of some of its terminally ill fellow citizens to use experimental agents, they might also remain unwilling to contribute financially to this behavior. Moreover, according to the previous argument, they should not be forced to support such actions by giving money to the acting person.

In regard to the costs described in point (1), I would like to refer to the summary of the preceding chapter. There I suggested that it is also in the best interests of society to allow people with terminal diseases to access experimental agents. I argued that this would be the case, as long as this takes place in a controlled way and as long as the experiences these people have are shared with other people with AIDS and the medical research community.

In regard to the costs addressed in point (2), the following argument, made by Norman Daniels, seems to be persuasive:

> Many lifestyle choices bring with them risks to health of greater magnitude than risks from many workplace health hazards. Yet people are not prohibited from smoking, drinking excessive alcohol, failing to exercise, or eating too much fatty meat. Nor are they prohibited from hang-gliding, scuba-diving, driving without seat belts [in the United States], or sun bathing.[29]

Given that we are willing to pay the health care bill for persons who have injured themselves simply to enjoy the thrill of a certain risk or to enjoy cigarette smoking, it would be quite inconsistent with this tradition if we were unwilling to pay the health care bill produced by persons who have taken these risks in order to survive an otherwise terminal illness. In fact, the very behavior that led people to develop sexually acquired AIDS (excluding cases of rape) was voluntary and based on informed consent. If we as a society concede people the right to accept risks as high as this, then I cannot see a plausible reason why we should be entitled to deny them the opportunity to take another high risk to try to save their lives.

ACCESS TO EXPERIMENTAL DRUGS IN SOCIETIES WITHOUT UNIVERSAL HEALTH CARE

In the Introduction, I alluded to some brief arguments in favor of universal health care. The purpose of this section is not to debate whether we should have universal health care, even in those democratic societies that seem to have rejected this idea. The only question to be discussed here is to what extent people with AIDS should be supported in their attempts to access experimental agents. In order to do this, I will assume that societies without universal health care do exist.

External constraints can easily render a person nonautonomous. This is particularly true for the poor and/or less educated people who live in societies without universal health care. The moral entitlement, or the legal right to buy and use experimental drugs, would be practically meaningless for this group of persons because poor people do not possess the necessary means to buy these agents. They might also encounter enormous problems if they tried to gather and evaluate the technical information necessary for the use of some drugs, even if it was made available. In countries without universal health care, certain disadvantaged people cannot even afford to consult a physician in order to get an understanding of how a certain experimental agent might affect their lives. These problems, while certainly significant, are not impossible to solve, at least, as I suggested earlier, in those countries that have universal health care for all of their citizens.

Do we have good reasons to argue for governmental health care support in societies without universal health care? One conceivable reason that is clearly insufficient to substantiate such a claim is that AIDS is a catastrophic and normally terminal illness. This is because other terminal illnesses have already set precedents for how AIDS sufferers should be treated. If people with terminal cancer do not receive special support because they suffer a fatal illness then there is no reason to assume that people with AIDS should be treated differently.

However, as I have shown in the first chapter, there are many important differences between the average person with AIDS and the average person with cancer. Some implications of the fact that the vast majority of people with AIDS belong to classically discriminated-against societal groups do help to substantiate an argument for increased governmental support for those individuals. This argument has been put forward persuasively, first by Richard D. Mohr for gay men,[30] and later by Patricia Illingworth,[31] who extended it to IDUs. It should be mentioned in this context that Mohr is a conservative, gay, libertarian philosopher and, as such, is not in favor of "welfare state" systems of health care provision. He believes, however, that in the case of AIDS and gay men, an argument for support of day-to-day care (and, as a corollary of this, I would argue, also for experimental agents) can be made. The basic premise of both his and Illingworth's argument is that society has harmed gay men and IDUs and, thus, owes them some form of compensation. Mohr argues that even "though AIDS is a disease caught in a condition where one has put oneself at risk, nevertheless gay men in general ought in consequence of certain natural and cultural forces to be viewed as morally innocent in its contagion and spread."[32]

Mohr presents his case for gay men by pointing out that society has prevented them from forming successful working families. Legal recognition for gay long-term partnerships is the exception, rather than the rule, and it exists only in a number of liberal European countries. Furthermore, "in a society where discrimination against gays is widespread, it is unlikely that gay couples will flourish."[33] Hence, argues Mohr, societies that prevent gay people from forming families should be required to assist gay people in situations where a caring family would help them under ideal circumstances. This is what compensatory justice requires. His libertarian position requires him to argue along the following lines to reach the conclusion I have just mentioned:

> And so even in the absence of justifications for perfectly general government health-care plans, funding should nevertheless be provided for the care of those whose life chances in civil society have been permanently destroyed by natural catastrophes that elude the protection of civil defense—so that their dying might at least avoid unnecessary suffering, and the indignity to which this disease in particular tends to expose patients.[34]

The whole idea presented by Mohr obviously relies on Robert Nozick's entitlement theory, which maintains that "whether a distribution is just depends upon how it came about."[35] Nozick's basic idea in this context is this: "From each according to what he chooses to do, to each according to what he makes for himself [perhaps with the contracted aid of others] and what others choose to do for him and choose to give him of what they've been given previously [under this maxim] and haven't yet expended or transferred." Further simplified, "From each as they choose, to each as they are chosen."[36] Mohr argues that the state legislation in regard to homosexual relationships has prevented homosexual men from choosing to organize their lives in a manner according to their wants and that this has actually harmed these people and made them more vulnerable to the acquisition of AIDS. Hence, according to Mohr, homosexual men with AIDS can expect compensatory justice from the state.

As Beauchamp and Childress have pointed out in regard to libertarian theories, "Justice consists in the operation of just procedures [such as fair play], not in the production of just outcomes [such as equal distribution of resources]. . . . There are no welfare rights, and therefore no rights to health care or claims to health care based on justice."[37] Thus, a libertarian could indeed argue that homosexual men with AIDS have not had such fair play, for the reasons presented, and deserve what Mohr calls compensatory justice.

In particular, Mohr proposes to fund the day-to-day care of end-stage AIDS patients. Of course, in order to enhance the autonomy of gay men, it seems clear to me that should we accept Mohr's argument. I see no reason, however, but to limit such compensatory funding to day-to-day care. In fact, the choices these people have can easily be further increased by introducing a system that is based on health care vouchers and that allows people to trade these vouchers for services of their liking (such as standard clinical care, alternative treatments, etc.).

Illingworth misses the point of Mohr's argument in her critique when she asks rhetorically whether "the fat, the ugly and the disfigured" should also be compensated.[38] Mohr is not concerned with people who have been wronged by social prejudices, but with people who have legally been denied the opportunity to cultivate a family. Obviously, a government cannot be held responsible for the plight of people who might be unable to find a partner because they are aesthetically undesirable for whatever reason. It can, however, be held responsible for its own laws that prevent people from forming a family on the basis of their gender orientation. If a government did prevent ugly people from forming families (and from receiving the legal protection that usually comes with it), then these discriminated-against people would have a reason to claim compensation in cases where this discrimination has led to harm. It is important to note that this kind of argument does not rely on any kind of normative premise that might ultimately lead us to accept any kind of welfare state.[39]

Surprisingly, Illingworth presents a line of argument very similar to Mohr's. The difference, however, is that she is not criticizing the legal discrimination that prevented gay men from setting up families but, rather, holds that "they are owed something because society has pretty much restricted their sexual lifestyle options to one, the fast track. Because this turned out to be a hazardous way of conducting a sexual life they suffered extremely harmful consequences, the most harmful of which is AIDS."[40] She extends the reasoning employed by Mohr to IDUs. She points out, persuasively I believe, that this other group of people at high risk for AIDS has been wronged in ways comparable to those that Mohr has presented in the case of gay men.

IDUs have contracted AIDS in high numbers because of the governmental policy of prohibition. This policy has prevented these people from accessing clean needles. Needle sharing became a fairly widespread phenomenon and has contributed significantly to the much higher percentage of AIDS cases in countries with rigid legislation, such as the United States. Legalized access to narcotics would have virtually eliminated the risk of contracting AIDS in this group of the population. Hence, Illingworth points out: "By looking at the ways in which society may have contributed to high-risk behavior, I have moved in the direction of attributing moral responsibility for HIV/AIDS, at least for the majority of those who have the disease, to society."[41]

Illingworth has argued that it is "important to compensate gay men and IV-users in a way that will enhance their autonomy because they have already endured a decrease in autonomy [due to the state's discriminatory legislation]." She rejects, then, my preferred solution of vouchers because she believes that "by dictating the terms of compensation, namely health care, the voucher system fails to maximize options."[42] Given the limited scope of the questions that I set out to answer in this book, I will not discuss the question of whether this limitation is really unacceptable because it allegedly does not allow people with AIDS to decide to spend the compensation otherwise (e.g., for the Gay Men's Christian League or to the Cancer Research Foundation, as Illingworth imagines). It seems clear to me, however, that one could imagine a situation where people are given vouchers for their health care needs and perhaps compensation in other form (perhaps money, as Illingworth proposes).

There is another reason in support of this kind of compensation that is of both a pragmatic and ethical nature. Currently, in the United States, for instance, people with AIDS can access some experimental drugs that are under investigation in clinical trials per expanded-access programs. Those patients, however, who intend to use these drugs under expanded access have to pay for these drugs, or else they can try to get into a clinical trial. This means that patients who can afford it will use these drugs under the expanded-access program while poor(er) patients will be left with no choice other than to join a clinical trial.[43]

PROVISION OF INFORMATION ABOUT EXPERIMENTAL DRUGS

I have argued that situations in which a person with AIDS has more, rather than less, information about a certain agent are preferable because they allow such a person to make his or her decision on a higher plane of first-level information. The premise on which this view is based is, of course, that this technical information would be of some use to a person with AIDS. For many highly educated people with AIDS, who have a lot of leisure time, information of such a technical nature would (all other things being equal) indeed improve the quality of their decision making.

However, imagine for a moment a situation in the United States that might be not all that fictitious. For the illegal IDU Latino immigrant, the existence, and the free availability of such information about a certain drug regimen, would be useless, even if he or she could afford to buy this experimental agent.

There are obviously all sorts of reasons why such a person with AIDS might be unable to study the available information carefully and fruitfully, even if it was available. In such situation, I think it would be of utmost importance to bring to this person's attention information about those drugs whose use is not indicated in AIDS therapy, especially if they are known to cause harm. It may well be that he or she has not the time available to balance the existing evidence for or against certain drugs, which might have both positive and negative effects. Taking these factors into account, it seems advisable (at the very least) to try to prevent him or her from using drugs that are undoubtedly harmful by providing technical information about substances or courses of treatment that are known not to work or that are known to have a negative clinical effect overall.

The more commonly invoked situation is one in which a white middle-class man with AIDS goes to his physician or another medical consultant who is likely to be an AIDS specialist. They can discuss the pros and cons of applying a new experimental agent on the basis of the available information. But, to return to our scenario, where can the Latino person go? There is nobody available to help him or her. In a good society, we would have to take care that not only all possible information (negative and positive) is made available to interested parties, but also, as is equally important, we would have to ensure that they are given the necessary technical knowledge to evaluate this evidence. Such a claim implies that specialists should provide free or affordable advice for people with AIDS who want it. The egalitarian ethical principle of equal respect and equal treatment of equal interests would require the provision of such a service.

On the other hand, in many Western countries that offer universal health care to their citizens, such as Australia, the overwhelming majority of people with AIDS are indeed gay, white, and male.[44] However, even under the much better circumstances in Australia, some people with AIDS face similarly insurmountable difficulties that are comparable to the problems with which poor, uneducated, ethnic minority members struggle in the United States. One such group is composed of all those gay men and IDUs with AIDS that come from non-English-speaking backgrounds. Vital information is translated into a limited number of languages, and even in this translated information the depth of information covered is restricted. Thus, only inadequate information is generally made available to people for whom English is not a native language. In neither a legal nor moral sense are these people incompetent, yet they all encounter specific problems when they attempt to gather vital information about alternative avenues to pursue in their search for a life-saving agent. These problems bring them to a much worse starting position than that of their white, middle-class fellow patients.

In Germany, none of these problems exist because this country has no immigrant population from non-German-speaking countries that accounts for an appreciable percentage of the overall number of AIDS cases. Hence, we might say that Germany is better off than the United States and even Australia in terms of the ease with which it can treat AIDS, insofar as the populations at risk for AIDS-defining diseases are relatively homogeneous, and universal health care is provided to all citizens at excellent levels.

Despite this advantage, however, the problem of different educational backgrounds of people with AIDS still needs to be overcome. A person with AIDS who has a university education in the biomedical sciences and the ability to efficiently search biomedical and other relevant databases is clearly in a much better situation than the AIDS patient who has only a high school education. The former will not only be able to gather first-level information efficiently, but also will be able to evaluate it in meaningful ways, which is equally important. Those people, however, who were not lucky enough to receive such an education, or who might even lack the intellectual competence to gather and analyze such information, are undoubtedly disadvantaged.

A feasible solution to these problems seems to be the creation of publicly and freely accessible databases that offer files for perusal by patients. Such files might include information about alternative therapies in addition to information about standard treatments.[45] They ought to include critical commentaries by proponents and opponents of a certain course of action.

There is something else that is to be said in regard to information about the impact of certain drugs on the quality of life of persons living with AIDS. Many studies that address this issue have been published in recent years. I have already mentioned the study by Lenderking and associates, which concluded that it is up to the patients to make the decision whether or not to use AZT. They came to this conclusion because in their judgment, although AZT did lead to a longer life, the normal advantages associated with an increase in longevity were canceled by the severe toxicity and side effects of this medication.

Such studies are extremely important because they can help patients make decisions about exactly which side effects they are willing to accept (loss of quality of life) in order to achieve certain gains (a longer life). As Carlton Hogan pointed out, "It is important to acknowledge that much of such evaluation is personal: Some people with AIDS might accept neuropathy, for example to escape dementia, or bone marrow suppression to avoid blindness due to CMV. These sorts of decisions are impossible without having information on clinical outcome."[46] It is crucial that the information provided to people with AIDS does not try to make these decisions on behalf of the patients. This would be a very significant limitation of their dispositional autonomy. In this context, it is worth recalling that I have mentioned numerous cases in which well-intentioned community-based AIDS organizations in Australia have attempted to make these decisions on behalf of people at risk for AIDS elsewhere.[47]

Many of this book's practical proposals are based upon the premise of an existing, ideal society that willingly sponsors certain projects in order to enhance the decision-making capabilities of some of its more vulnerable citizens. Unfortunately, such societies do not currently exist. Health care funds are scarce resources, and we have to think twice before spending such funds. I think, therefore, that we should put a high priority on making information regarding experimental drugs available to people with AIDS in order to prevent them from harming themselves by using those agents that are undoubtedly harmful overall.[48]

A good practical project that is going in the right direction is a recent initiative by the U.S. National Library of Medicine (NLM). It announced the "solicitation of quotations from community-based organizations, patient advocacy groups, and libraries to design and conduct local programs for improving AIDS-related information access for patients and the affected community as well as their care givers."[49] Acknowledging different levels of education in the affected communities, the National Library of Medicine emphasizes "providing AIDS-related information or access to AIDS-related information in a way meaningful to the target community."[50]

Another project, again undertaken in the United States, also seems to be laudable in terms of its broad-minded approach. The National Institutes of Health (NIH) have given up their principled rejection of everything deemed alternative medicine. Their Office of Alternative Medicine, which, in 1995, boasted funding of $5.4 million (out of $10 billion for the NIH's overall budget), has agreed to give an $840,000 research grant to Bastyr University in Seattle to "catalog and describe alternative therapies currently used for HIV and AIDS patients and to evaluate the effectiveness of the more promising ones."[51]

A number of reasons might provide the motivation for this change of attitude among mainstream medical institutions such as the U.S. NIH and the NLM. One primary reason is the failure of mainstream medicine to produce a working therapy for AIDS. Robert Root Bernstein, professor of physiology at Michigan State University, has described other reasons:

> First, . . . we must consider the possibility that some nontraditional approaches to AIDS may, beneath their irrational veneers, also contain viable nutritional, herbal, and behavioral approaches to disease—perhaps even chemical leads to pharmacologically active compounds—that science needs to tease out. . . . Second, physicians have been loath to accept even well-documented treatment modalities such as hypnosis and acupuncture because the underlying phenomena have no rational explanation within our current theoretical framework. Alternative hypotheses to AIDS provide alternative frameworks for interpreting as valid some otherwise unexplainable treatments or "cures" . . . And finally, we must always remember when we evaluate novelties in science that the greatest medical breakthroughs . . . were all met . . . with total skepticism by the medical community. I would not be surprised if the most important innovators in AIDS research and treatment turn out to be peripheral members of the research and treatment communities.[52]

So far, I have not made any distinction in regard to the specifics of the provision of information between societies with universal health care and those without. Obviously, however, there are significant differences between the sets of values accepted in these respective societies on which to base an argument. It is clearly easier to argue for the financing of the provision of this information in societies that provide universal health coverage than it is for those societies that do not.

Provision of Information in Societies with Universal Health Care

In Germany, universal health care is provided and the consumer (patient) is usually well informed about the possible side effects of a given drug that his or her physician intends to prescribe. The drug packages themselves contain lengthy information about the available knowledge regarding the prescribed drugs, including side effects and possible contraindications. The content of the information is strictly regulated by governmental agencies and does not depend upon the goodwill of the pharmaceutical company that is producing the drug or upon the information the physician is willing to provide.

The producers of experimental agents can reasonably be required to provide to the consumer (patient) the knowledge they have under headings such as dosage, side effects, efficacy, etc. If there is no such information available, then, in consistence with the argument that I have provided in the chapter on paternalism, I propose that the manufacturer is required to write just that into the respective entries to make sure that the patient is not misinformed about the degree to which the available knowledge is limited. Also, given that the drug is not governmentally approved, the producer of the agent should be required to inform the prospective user about how the knowledge that has been provided has been gathered. If, for instance, no (pre-)clinical trials have taken place, then this information should be conveyed to the consumer. Undoubtedly, such provision of information would have to be tightly controlled by a regulatory governmental agency. On the other hand, we would have to be able to ensure that this agency does not use its powers in order to prevent terminally ill people from accessing experimental agents.

Provision of Information in Societies Without Universal Health Care

What can we say about the situation in societies without universal health coverage? First of all, it should be pointed out that in most parts of the developed world, strict rules are in place in regard to the information the pharmaceutical industry has to supply to patients. At the very least, one would expect something similar to be required, even under conditions of nonexistent universal health care. But this is, unfortunately, where the obligation of the state seems to cease. Whereas in countries with universal health care, the state covers the costs that are incurred when patients consult physicians, in countries such as the United States, this is not normally the case. For this reason, the aforementioned proposals to allow patients to access information via information (data-) banks seems most practical. Such (minimal) suggestions are, of course, not of the same quality as a consultation with a qualified physician. Obviously, a situation that gives patients the opportunity to discuss the available information about an experimental agent with a physician of their choice would increase the degree to which they are informed, far more than would interaction with a machine. Hence, their dispositional autonomy would be expected to be stronger than that of people living under a system that does not guarantee universal health care.

How might countries without universal health coverage be provided with reasons to accept my proposal of a freely accessible database? One could reasonably argue that perhaps some sort of contract between patients, the state, and the pharmaceutical industry could lead both to an improvement of the patients' baseline situation and to benefits for all involved. This is because carefully monitored use of experimental agents by people with AIDS, even though it cannot substitute for a carefully controlled clinical trial, leads to better information about the impact of a certain drug. Undoubtedly, this is not only in the interest of the patient, but also of prospective future patients and the state, which needs this information in order to make decisions in regard to which of the experimental drugs should be tried in the existing clinical trial protocols.

Finally, even the pharmaceutical industry has an interest in gathering this knowledge because, should their drug have a demonstrably positive impact on patient survival and/or quality of life, they would be better able to market it to larger numbers of patients. For these reasons, I find it intelligible to argue in favor of some sort of contract between patients, the governments' regulatory bodies, and the pharmaceutical industry that leads to a solid financial basis for such a database.

PRACTICAL CULTURAL HINDRANCES IN EXERCISING AUTONOMY

As much as any similarly situated group of patients, people with AIDS experience internal obstacles in their attempts to exercise autonomy. This has been highlighted in a recently published study. The research suggests that the reasons for empirically verifiable differences in terms of prophylaxis of certain opportunistic infections between whites and people of ethnic minorities are not necessarily due to economic and educational differences, as one might initially assume. This study (published in 1994, in the *New England Journal of Medicine*) shows that significant differences in the use of prophylactics against *Pneumocystis carinii* pneumonia exist among people with AIDS, depending on whether they are white or members of ethnic minorities. This finding has especially important public health implications, as AIDS has disproportionately affected racial and ethnic minorities in the United States. Specifically, the study found that 82 percent of eligible whites, but only 58 percent of eligible blacks, received PCP prophylactics. It was concluded that these differences did not correlate with factors such as age, sex, type of insurance, income, or education. Seemingly "this disparity suggests a need for culturally specific interventions to ensure uniform access to care."[53]

Indeed, similar studies have been undertaken in the United States among patients with heart problems and among 10,000 patients attending hospitals in five states, all of which came to conclusions similar to the aforementioned study of AIDS patients.[54]

Putative internalized cultural hindrances, which lead to the choosing of a less than optimal course of action by a given person with AIDS concerning their treatment decisions, have been characterized by Young as internal obstacles.[55] If they existed, they would need to be carefully addressed, if we are serious about increasing the decision-making capabilities of all people with AIDS. One way might be to engage in a dialogue with physicians who actually belong to such minority groups to get a better understanding of cultural (including religious) factors which lead to decisions that are undoubtedly less than ideal.

Conclusion

The conclusions of this analysis are certainly not in agreement with either the current state of the art or common sense in two important areas of our public health policies. First, I have argued that it is not possible to use the widely accepted standard theories of autonomy, and their corresponding theories allowing weak and strong paternalistic interventions, in order to justify preventing terminally ill people from buying and using experimental agents. The major motive for a paternalistic intervention is to prevent harm to self. The intervention we are concerned with in most cases of people with AIDS is not of a weak paternalistic nature but, rather, a strong one. The claim that these interventions are not weakly paternalistic derives from the simple fact that the patients in question are usually perfectly aware of the risk that they would take if they use an experimental agent.

I have argued that the harm done to a dying patient by preventing him or her from testing an experimental agent is ultimately of greater gravity than the harm caused by this drug itself, both in the case that it fails to work and in the more serious event that it has severe side effects. I supported this conclusion by emphasizing that it is a direct consequence of the nature of a terminal illness that people will lose their lives. Given that standard therapies do not work,[1] I suggested that it might be perfectly reasonable for a terminally ill patient to take his or her chances with an experimental agent about which very little is known. A dying person, in particular, should be entitled to decide how he or she will live the remainder of his or her life. We, both as a society, and as health care professionals, should respect their autonomous decisions. In the case of a terminal illness, such decisions may cover options ranging from doing nothing, to joining standard clinical trials, to using experimental agents, or to committing suicide (perhaps by asking for voluntary euthanasia).[2] A permanent intervention, such as a paternalistic action designed to prevent use of the desired drug, is too great an interference because it would prevent this person from doing what he or she wants to do in a situation that is of utmost importance for his or her life.

It is important to stress that the scope of my argument is limited to terminally ill people. The fact of terminal illness is the crucial basis of the particular argument I have suggested, which aims to prevent certain strong paternalistic interventions of the Youngian type. Indeed, in most cases concerning patients with a curable illness, I consider Young's justification for strong paternalistic interventions persuasive.

The second part of this book has argued that the current situation of nonaccess to experimental therapies effectively forces terminally ill people, by default, to join clinical trials to obtain the drug(s) they want to use. The major ethical justification that is expressed by proponents of this situation is not based on paternalistic motives, but on arguments related to harm to others. Its defenders have suggested that, without the participation of current patient generations in clinical trials, there will be no drugs for future patients. It is for this reason that they think it is acceptable to coerce current patients into the participation in clinical trials.

I have rejected this argument, mainly on the grounds that it defends a situation in which a terminally ill patient faces a scenario comparable to that of passengers in a hijacked aircraft. In the latter case, someone holds a gun to the passenger's head and asks him or her to make a choice between dying or adhering to the rules set by the hijacker.[3] Similarly, terminally ill patients are currently left with the options of not getting access to experimental agents except, perhaps, by joining a clinical trial that tests the desired drug or by ending up with a standard treatment that is, by definition, unsuccessful (where the criterion for success is prevention of AIDS-related death).

These clinical trials are often designed in ways that violate patient survival interests and/or exclude large parts of the patient population. Patients have realized this and have subsequently started cheating in order to meet the entry requirements of clinical trials. They may also cheat within the trial, perhaps by engaging in the (forbidden) use of concomitant drugs. Obviously, most patients are unable to act altruistically to the extreme that requires them to put their individual survival interests behind the research interests (and the assumed interest of future patient generations) of a given trial. Rather, most terminally ill patients go to great lengths to save their lives. I have argued that, if we cannot expect to find patients who would truly volunteer to act in such an altruistic manner, then we must also assume that there cannot be a democratic consensus in which the members of a given society agree to sacrifice themselves for the sake of future patients. Moreover, I have claimed that, under the current circumstances, where patients have no real choice other than to join a clinical trial, we cannot reasonably assume that they are truly volunteers. Hence, I have concluded that they have no reason to adhere to trial requirements that require them to sacrifice vital survival interests, simply because they are not volunteers in the first place. They have not given autonomous consent to be subjects in such a trial.

Furthermore, I showed that the current research designs, which prevent large subpopulations of the overall patient population from joining clinical trials, and which often have a long list of prohibitions in regard to the concomitant use of other drugs, have no real value for prospective patients. This is the case because in the real world (as opposed to the situation in a clinical trial), patients do not receive substandard treatments, at least in developed countries with universal health care. If a trial proscribes such substandard treatments by excluding widely used drugs, its results are only of value for those few who do not use the prohibited drugs. In other words, the current drug approval system, combined with the clinical trials system as we know it in virtually all Western countries, not only violates the individual autonomy of terminally ill patients in an unacceptable manner, it also leads to methodologically questionable results. The consequence of this situation is harm to future patient generations caused by unreliable trial results.

Thus, the second part of this book argues that we are not ethically entitled to prevent terminally ill patients from accessing experimental drugs in order to develop new therapeutic agents in standard clinical trials. I suggest that people with terminal illnesses should be given permission to access experimental agents of their liking, as long as they are aware of the risks such action entails. Physicians should administer the prescription of these agents and should monitor the effect such drugs have on patients.

In the last chapter of this book, I have proposed that the costs that might occur when we give terminally ill patients suffering AIDS-defining illnesses access to experimental agents should be covered by the public health care system in countries with universal health care. This coverage should extend, at least, to the amount that is normally paid in these countries for the (unsuccessful) standard treatments that are available. Furthermore, I suggested that, in societies without universal health care, there is also good prima facie reason to reimburse people with AIDS for the costs they incur. This reason derives from the recognition that their suffering from AIDS is, to a significant degree, caused by a society that has created living conditions for these minority groups which have dramatically increased their chances of contracting AIDS. Finally, I have proposed that, as a society, we should assist terminally ill people in making as informed a decision as possible about any experimental therapies that they might wish to use. In particular, the systematic monitoring of the experiences people with AIDS have while they use experimental drugs should be gathered through their physicians and evaluated by suitable personnel.

In summary, people with AIDS or other terminally ill patients should be given access to experimental drugs. The access should be provided through qualified physicians, and the results of the patients' experiences should be carefully monitored and systematically evaluated. The current situation, which legally prevents terminally ill patients from accessing experimental agents and forces them, by default, to join clinical trials, is ethically unacceptable. We cannot expect such patients to act altruistically to the extent that is currently required from participants in clinical trials. Independently of this claim about what patients can and cannot do, research clinical trials have an ethical obligation to try to accommodate the participating patients' survival interests more than they have in the past and more than many do today. Finally, the costs incurred to health care systems by patients who choose to buy and use experimental drugs should, to some degree, be covered by the state.

Notes

Introduction

1. See, for example, the cover photo of Crimp, D. (Ed.). *AIDS: Cultural Analysis, Cultural Activism*. Cambridge, MA: MIT Press, 1988.

2. For a good account of the history of related activism see Nussbaum. *Good Intentions: How Big Business and the Medical Establishment Are Corrupting the Fight Against AIDS, Alzheimer's, Cancer and More*. New York: Penguin, 1990.

3. Altman, D. *AIDS and the New Puritanism*. London: Pluto Press, 1986: 33.

4. Crimp, D. AIDS: Cultural Analysis/Cultural Activism. In Crimp, D. (Ed.). *AIDS*, 1988: 8.

5. See photo of an ACT UP demonstration at the Memorial Sloan-Kettering Hospital in New York City, July 21, 1987. Demonstrators carried a placard with the text, "5000 PWAs in New York: only 149 being in current drug trials." In Crimp, D. (Ed.). *AIDS*, 1988: 152.

6. Sonnabend, J. Do We Need New Ways to Evaluate Experimental AIDS Treatments? *AIDS FORUM*, 1989; 1(1): 6. Emphasis added in the first case of italics in this quote but not in the second.

7. Edgar, H. and Rothman, D. J. New Rules for New Drugs: The Challenge of AIDS to the Regulatory Process. *Milbank Quarterly*, 1990; 68(1): 113. (Subsequently quoted as New Rules for New Drugs.)

8. Edgar, H. and Rothman, D. J. New Rules for New Drugs, 1990: 113.

9. Hodel, D. FDA Reform Floated in DC. GMHC *Treatment Issues*, 1995; 9(6): 6-7.

10. Novick, A. Reflections on a Term of Public Service with the FDA Anti-Virals Advisory Committee. *Aids and Public Policy Journal*, 1993; 8(2): 55. (Subsequently quoted as Reflections.)

11. The AIDS epidemic has led the United States to introduce some mechanisms that allow this legislation to be circumvented in exceptional circumstances. It is possible, for example, for a physician to "administer an unapproved new drug when a patient has not responded to approved therapies." See Flannery, E. J. Should It Be Easier or Harder to Use Unapproved Drugs and Devices? *Hastings Center Report*, 1986; 16(1):19. However, there are additional hurdles that make it impossible for many patients and physicians to choose this avenue. For instance, it is necessary that the physician develops a clinical trial with that drug and that he or she attains the approval of the FDA and an Institutional Review Board.

12. A prospective patient is a person who will, in the case of AIDS for example, acquire this disease in the future and who might then die unless success-

ful therapies have been developed and are accessible for everyone who needs them.

13. See, for example, Nussbaum, B. *Good Intentions*, 1990; Kwitny, J. *Acceptable Risks*. New York: Poseidon Press, 1992.

14. National Research Council. *Social Impact of AIDS*. Washington DC: National Academy Press, 1993.

15. As quoted in Kwitny, J. *Acceptable Risks*, 1992: 10.

Chapter 1

1. I link discrimination on the basis of sex to discrimination on the basis of sexual orientation because it seems to me that there is reason to believe that the latter is really a particular type of the former discrimination. If sexual partners are of the opposite sex, there is no discrimination on the basis of sexual orientation. Two men, or two women, however, can be killed, tortured, or subjected to involuntary psychiatric treatment in many parts of the world for committing the same sexual acts for which a man and a woman would not be punished. Hence, it is fairly obvious that this kind of discrimination is primarily based upon sex. Schüklenk, U. and Riley, T. Homosexuality, Societal Attitudes Toward. In Chadwick, R. (Ed.). *Encyclopedia of Applied Ethics*. Volume 2. San Diego, CA: Academic Press: 597-603.

2. Ironically, the Northern Territory has recently legislated that voluntary euthanasia will become legal in this part of Australia. In other words, suicide will be legal when it complies with this legislation (i.e., the person who plans to commit suicide must suffer a terminal illness).

3. Beauchamp, T. L. and Childress, J. F. *Principles of Biomedical Ethics*, Fourth Edition [subsequently quoted as *Principles*]. New York: Oxford University Press, 1994: 120.

4. Young, R. *Personal Autonomy: Beyond Negative and Positive Liberty* [subsequently quoted as *Personal Autonomy*]. Kent: Croom Helm, 1986: 1.

5. Mohr, R. D. AIDS, Gays, and State Coercion. *Bioethics,* 1987; 1(1): 35-50.

6. Bayer, R. *Private Acts, Social Consequences. AIDS and the Politics of Public Health*. New Brunswick, NJ: Rutgers University Press, 1991.

7. Illingworth, P. *AIDS and the Good Society*. London: Routledge, 1990.

8. Mill, J. S. *On Liberty*. Included in *Utilitarianism, Liberty, Representative Government* [subsequently quoted as *Liberty*]. London: Dent, 1960.

9. Mill, J. S. *Liberty*, 1960: 73.

10. Mill, J. S. *Liberty*, 1960: 117.

11. Feinberg, J. *The Moral Limits of the Criminal Law: Harm to Self* [subsequently quoted as *Harm to Self*]. New York: Oxford University Press, 1986: 134-142.

12. Schüklenk, U. HIV-Related Phraseology [letter]. *Australian Journal of Public Health*, 1992; 16(3): 329-330.

13. National Research Council. *Social Impact of AIDS*. Washington, DC: National Academy Press, 1993.

14. *Berita AIDS Pusat Perubatan.* Newsletter published by University of Malaya. 1993; 1(1): 3.

15. Schüklenk, U., Mertz, D., and Richters, J. The Bioethics Tabloids: How Professional Ethicists Have Fallen for the Myth of Tertiary Transmitted Heterosexual AIDS. *Health Care Analysis,* 1995; 3(1): 27-36.

16. Bayer, R. *Homosexuality and American Psychiatry,* Second Edition. Princeton: Princeton University Press, 1987. Indeed, organizations such as the World Islamic Mental Health Association still consider homosexuality "a mental disorder and feel these people require treatment and management by the mental health professionals." (Prof. Muhammad Rashid Chaudhry, President, World Islamic Mental Health Association, letter, September 22, 1997.)

17. Schüklenk, U. Naturwissenschaften und Philosophie. Lautmann, R. (Ed.). *Homosexualität. Handbuch der Theorie- und Forschungsgeschichte* [subsequently quoted as Naturwissenschaften]. Frankfurt am Main, Germany: Campus Verlag, 1993: 307-312. See also Schüklenk, U. and Ristow, M. The Ethics of Research into the Causes of Homosexuality. *Journal of Homosexuality,* 1996; 31(3/4): 5-30.

18. I will restrict the subsequent analysis to the situation of gay and bisexual men. I will mention the situation of IDUs only in cases in which I believe there to be ethically relevant differences.

19. National Gay and Lesbian Task Force Policy Institute. *Anti-Gay/Lesbian Violence, Victimization, and Defamation in 1991.* Washington, DC: NGLTF, 1992.

20. Herek, G. M. Hate Crimes Against Lesbians and Gay Men—Issues for Research and Policy. *American Psychologist,* 1989; 44: 948-955.

21. Young, R. *Personal Autonomy,* 1986: 21.

22. Schüklenk, U. and Mertz, D. Christliche Kirchen und AIDS. In Dahl, E. (Ed.). *Die Lehre des Unheils. Fundamentalkritik am Christentum.* Hamburg, Germany: Carlsen Verlag, 1993; 263-279, 309-312.

23. Kleinig, J. *Paternalism.* Totowa, NJ: Rowman and Allanheld, 1984: 14.

24. Kant, I. *The Metaphysics of Morals* [subsequently quoted as *Metaphysics*]. Cambridge, MA: Cambridge University Press, 1991: 64.

25. Kant, I. *Groundwork of the Metaphysics of Morals* [subsequently quoted as *Groundwork*]. Translation by H. J. Paton. New York: Harper & Row, 1964.

26. Kant, I. *Critique of Practical Reason* [subsequently quoted as *Critique*]. Translation by L. W. Beck. New York: Garland, 1976: 191. ("Die Gesinnung, die ihm dieses zu befolgen obliegt, ist: aus Pflicht, nicht aus freiwilliger Zuneigung und auch allenfalls unbefohlener, von selbst gern unternommener Bestrebung zu befolgen." Kant, I. *Kritik der Praktischen Vernunft.* Hamburg, Germany: Felix Meiner Verlag, 1929: 99.)

27. Kant, I. *Groundwork,* 1964: 57.

28. Kant, I. *Groundwork,* 1964: 92.

29. Schopenhauer, A. *On the Basis of Morality* [quoted from the translation by E. F. J. Payne]. Indianapolis, IN: Bobbs-Merrill, 1965: 62.

30. Kant, I. *Groundwork,* 1964: 80.

31. Kant, I. *Groundwork,* 1964: 70.

32. Kant, I. *Groundwork*, 1964: 88.
33. Kant, I. *Groundwork*, 1964: 96.
34. Kant, I. *Critique*, 1976: 67-68.
35. Kant, I. *Critique*, 1976: 123-124.
36. Kant, I. *Groundwork*, 1964: 69.
37. The following passage also seems to support this interpretation: "But man regarded as a person, that is, as a subject of morally practical reason, is exalted above any price; for as a person (*homo noumenon*) he is not valued merely as a means to the ends of others or even to his own ends, but as an end in itself, that is, he possesses a dignity (an absolute inner worth) by which he exacts respect for himself from all other rational beings in the world." Kant, I. *Metaphysics*, 1991: 230.
38. Kant, I. *Metaphysics*, 1991: 50.
39. Kant, I. *Metaphysics*, 1991: 56.
40. Mill, J. S. Utilitarianism. In Mill, J. S. *Liberty*, 1960: 49.
41. Kant, I. *Groundwork*, 1964: 96. ("Act in such a way that you always treat humanity, whether in your own person or in the person of any other, never simply as a means, but always at the same time as an end" [p. 96].)
42. Hare, R. M. Could Kant Have Been a Utilitarian? [subsequently quoted as Kant]. *Utilitas*, 1993; 5(1): 2.
43. Kant, I. *The Doctrine of Virtue* [subsequently quoted as *Virtue*]. Translated by Mary J. Gregor. New York: Harper & Row, 1964: 117. See also Kant, I. *Groundwork*, 1964: 98. ("For the ends of a subject who is an end in himself must, if this conception is to have its full effect in me, be also, as far as possible, my ends" [p. 98])
44. Hare, R. M. Kant, 1993: 4.
45. Smart, J. J. C. An Outline of a System of Utilitarian Ethics. In Smart, J. J. C. and Williams, B. (Eds.). *Utilitarianism: For and Against*. Cambridge, MA: Cambridge University Press, 1973: 9.
46. Paton, H. J. Analysis of the Argument. In Kant, I. *Groundwork*, 1960: 33.
47. Schopenhauer, A. *On the Basis of Morality*, 1965.
48. Kant, I. *Lectures on Ethics* [subsequently quoted as *Lectures*]. Translated by L. Infield. London: Methuen, 1930: 148-154, 169-171. Kant argues here, for instance on the subject of homosexuality, that deeds he describes under the heading *crimina carnis contra naturam* are "against the ends of humanity." In good Christian tradition, he assumes that the ends of humanity lie ultimately in procreation.
49. Kant, I. *Lectures*, 1930: 149. See also Kant, I. *Groundwork*, 1964: 89.
50. Kant, I. *Lectures*, 1930: 152.
51. Kant, I. *Lectures*, 1930: 152.
52. Hare, R. Kant. 1993: 5.
53. Schopenhauer, A. *On the Basis of Morality*, 1965: 120.
54. Schopenhauer, A. *Preisschrift über das Fundament der Moral*. Hamburg, Germany: Felix Meiner Verlag, 1979: 84. ("Ich muß daher den Ethikern den paradoxen Rath erteilen, sich erst ein wenig im Menschenleben umzusehen" [p. 84].)

55. Mill, J. S. *Liberty*, 1960: 75.
56. Mill, J. S. *Liberty*, 1960: 73.
57. Aristotle. *Nicomachean Ethics*. V.II.1138a12 [quoted from the translation by David Ross]. Oxford: Oxford University Press, 1966: 134.
58. Schopenhauer, A. *On the Basis of Morality*, 1965: 58. (In his translation, E. F. J. Payne used the word "injury" instead of "injustice." Given that the original Schopenhauer, Arthur. *Preisschrift über das Fundament der Moral*. Hamburg: Felix Meiner Verlag, 1979: 24 uses the word *Unrecht*, I have replaced injury with injustice partly because injury makes no sense in this particular context. More important, injustice is what Schopenhauer used in the German language version: "Da nämlich Das, was ich thue, allemal Das ist, was ich will; so geschieht mir von mir selbst auch stets nur was ich will, folglich nie Unrecht."
59. Mill would therefore reject legal moralism of the type Patrick Devlin has proposed. See Devlin, P. *The Enforcement of Morals*. London: Oxford University Press, 1965. Lord Devlin argued that the legalization of homosexual acts constitutes de facto public harm because it breaks some sort of (minimal) societal moral consensus that is fundamental for the existence of any society. As such, it might threaten the very fabric of such a society. He holds the view that every society has the right to defend itself against such assaults and is therefore entitled to enforce legally the moral views of the majority (pp. 11ff and 16ff). For critical responses see: Dworkin, R. *Taking Rights Seriously*. London: Duckworth, 1978 (Chapter 10); Hart, H. L. A. *Law, Liberty and Morality*. Oxford: Oxford University Press, 1963: 13-25, 67-68; Wollheim, R. Crime, Sin, and Mr. Justice Devlin. *Encounter*, 1959; 13: 34-40.
60. Mill, J. S. *Liberty*, 1960: 133.
61. Mill, J. S. *Liberty*, 1960: 132-133.
62. Mill, J. S. *Liberty*, 1960: 135.
63. I think that it is logically possible that I want the possible harm in a way similar to my wanting the benefits, when I take a certain risk. If I am an autonomously acting individual, and I am not willing to take my chances in a given risky operation, it is inconceivable that I would (voluntarily) agree to take the risk in the first place. For a comparable line of argument see Faden, R. R. and Beauchamp, T. L. *A History and Theory of Informed Consent* [subsequently quoted as *HTIC*]. New York: Oxford University Press, 1986: 244-247.
64. Mill, J. S. *Liberty*, 1960: 137.
65. Mill, J. S. *Principles of Political Economy*. Book I and V [subsequently quoted as *Political Economy*]. Harmondsworth, England: Penguin, 1970: 322.
66. Mill, J. S. *Liberty*, 1960: 152.
67. Mill, J. S. *Liberty*, 1960: 73.
68. Shorr, A. F. AIDS and the FDA: An Ethical Case for Limiting Patient Access to New Medical Therapies. *IRB: A Review of Human Subjects Research*, 1992; 14(4): 2.
69. The only exceptions to this rule, as far as I am aware, are children. They are considered incompetent to make legally binding decisions qua being a member of the group of underage persons.

70. One exception, perhaps, could be a hastened disease progression, caused by severe side effects of the experimental agent or therapy. But even in this case, I cannot see that the risk the patient is running is much higher than that of persons who join standard clinical trials. As I will demonstrate in the next chapter, even in standard clinical trials, survival interests of patients are often neglected, and they frequently receive substandard treatment. However, even if the risk is higher in the case of the unapproved drug, it is far from evident that to take this risk is unreasonable to a degree that proves that the person in question is incompetent.

71. On the level of personal experiences that I have had with persons with AIDS, I would contend that most of those whom I have had the opportunity to get to know balanced the possible pros and cons of the existing options both carefully and rationally. After this often painful deliberation, emphasized by the fact that none of the options are optimal, they variously decide to join a clinical trial, try a new unapproved drug, change their course of therapy, stick to the course of treatment they are currently on, etc. However, I am also prepared to admit that the panicking individuals Shorr has painted in his article might exist as well. What is unacceptable, though, is to employ quasi-inductionist reasoning and to claim that, therefore, all members (or even a majority) of this group are unable to make rational, substantially autonomous decisions.

72. Buchanan, A. E. and Brock, D. W. Deciding for Others. *Milbank Quarterly*, 1986; 64(2): 34-35. See also their *Deciding for Others: The Ethics of Surrogate Decision Making* [subsequently quoted as *Surrogate Ethics*]. Cambridge: Cambridge University Press, 1989.

73. Mill himself sought to change the British law in this respect because he believed that it was too easy to declare citizens incompetent. See Mill, J. S. *Political Economy*, 1970: 322.

74. Mill, J. S. *Liberty*, 1960: 73. "We are not speaking of children, or of young persons below the age which the law may fix as that of manhood or womanhood. Those who are still in a state to require being taken care of by others, must be protected against their own actions as well as against external injury" (p. 73).

75. Beauchamp, T. L. and Childress, J. F. *Principles*, 1994: 123. See also Faden, R. R., Beauchamp, T. L. *HTIC*, 1986: 237-241.

76. Mill, J. S. *Liberty*, 1960: 69.

77. Mill, J. S. *Liberty*, 1960: 74.

78. Ten, C. L. *Mill on Liberty*. Oxford: Oxford University Press, 1980: 41.

79. Harris, R. N. Private Consensual Adult Behavior: The Requirement of Harm to Others in the Enforcement of Morality. *UCLA Law Review*, 1967: 585n.

80. Beats is a slang term describing places where gay men can have casual anonymous sex, including venues such as public toilets or parks. Sullivan must not be fully aware of the meaning of the word beats in gay subcultures, otherwise she would not have seriously proposed to close them. After all, it is clearly practically impossible to close down all public places where gay men might come together to have sexual encounters, lest we soon be left without parks, public toilets, streets, etc.

81. Sullivan, L. AIDS—Beyond Education. Letter to the editor. *Medical Journal of Australia*, 1992; 157: 286.
82. Chapman, S. AIDS—Beyond Education. Letter to the editor. *Medical Journal of Australia*, 1992; 157: 501.
83. Beauchamp, T. L. Paternalism and Bio-Behavioural Control. *The Monist*, 1976; 60(1): 78.
84. Mill, J. S. *Liberty*, 1960: 151-152.
85. Ten, C. L. *Mill on Liberty*, 1980: 109.
86. Mill, J. S. *Liberty*, 1960: 152.
87. Mill, J. S. *Liberty*, 1960: 158.
88. The incoherence is related to Mill's preceding view, which argued that the individual can basically do whatever he or she wants in regard to matters which affect only herself.
89. Ten, C. L. *Mill on Liberty*, 1980: 118.
90. Ten, C. L. *Mill on Liberty*, 1980: 118-119.
91. Regan, D. H. Justifications for Paternalism. In Pennock, J. R. et al. (Eds.). *The Limits of Law*. New York: Lieber-Atherton, 1974: 189-210.
92. Regan, D. H. Justifications for Paternalism, 1974: 205.
93. Regan, D. H. Justifications for Paternalism, 1974: 206.
94. Regan, D. H. Justifications for Paternalism, 1974: 206.
95. Derek Parfit thinks that we change our personal identities over time anyway. See Part Three of his *Reasons and Persons*. Oxford: Clarendon Press, 1989.
96. Mill, J. S. *Liberty*, 1960: 74.
97. Mill, J. S. *Liberty*, 1960: 73.
98. Arneson, R. J. Mill versus Paternalism. *Ethics*, 1980; 90(4): 470-489.
99. Komrad, M. S. A Defense of Medical Paternalism: Maximizing Patients' Autonomy. *Journal of Medical Ethics*, 1983; 9: 39.
100. Mill presents, though, a similar argument in his *Political Economy*, 1970: 325. He argues here again that "a second exception to the doctrine that individuals are the best judges of their own interests, is when an individual attempts to decide irrevocably now, what will be best for his interests at some future or distant time" (p. 325). Robert Young has interpreted this in his *Personal Autonomy* so that he concludes Mill "allows paternalistic interference . . . where an irrevocable decision has been taken which would have far-reaching consequences on the distant." 1986: 66. I think this is indeed what Mill had in mind when he wrote this sentence. However, it appears clear to me that this practical conclusion contradicts Mill's general approach too, which considers the individual as the master of his or her own life and prohibits any intervention in that life for the person's own sake. Additionally, this argument is, to a certain degree, practically useless because it applies to all possible actions that we wish to undertake. For any action we wish to undertake, be it going to the movies, having a dinner with a friend, taking unapproved drugs, or boarding an aircraft, we cannot possibly claim to know what its consequences might be. It could be the case that on the way to the cinema I have an accident that leaves me crippled for the rest of my life, the aircraft might crash, etc. I cannot possibly foresee what the consequences of most decisions in

my day-to-day life are. I might have a fairly realistic prima facie assumption of what will happen, and this may indeed, more often than not, happen, but undoubtedly, I cannot know for sure.

101. Dworkin, G. Paternalism. In Wasserstrom, R. A. (Ed.). *Morality and the Law.* Belmont, CA: Wadsworth Publishing, 1971: 118.

102. Dworkin, G. *The Theory and Practice of Autonomy* [subsequently quoted as *Theory and Practice*]. Cambridge, MA: Cambridge University Press, 1988, p. 129.

103. Among other arguments, a similar point of view has been put forward by Joel Feinberg in his essay Legal Paternalism. *Canadian Journal of Philosophy,* 1971; 1(1): 118-119. Though he presents us also with another line of argument, claiming that "the legal machinery for testing voluntariness would be so cumbersome and expensive as to be impractical" (pp. 118-119), this argument is not so much of a moral than of a practical nature, and it has been convincingly rejected by Robert Young, who pointed out that we already test voluntariness in other important legal matters and find this useful and practically possible. See Young, R. *Personal Autonomy,* 1986: 75.

104. Say we assume that this would have no negative impact for the drug approval process as a whole and, therefore, no negative implications for future patient generations.

105. For instance, a disclaimer on the packages, similar to those on cigarette packets, might do. Of course, in many such situations the truth is that no information about either toxicity or efficacy of an experimental agent is available. However, in this situation, it seems sufficient to keep the potential risk taker informed of exactly that state of our knowledge.

106. Goodin, R. E. The Ethics of Smoking. *Ethics,* 1989; 99: 583.

107. Mill, J. S. *Liberty,* 1960: 152. H. L. A. Hart thinks that Mill clearly goes over the top here. "Mill carries his protest against paternalism to lengths that may now appear to us fanatic. . . . If we no longer sympathize with this criticism this is due . . . to a general decline in the belief that individuals know their own interest best." Hart, H. L. A. *Law, Liberty and Morality,* 1959: 32. There is certainly some truth in this argument as far as the general society is concerned. However, it almost certainly fails to convince in the case of many people with AIDS-defining diseases. They are often better informed than their physicians and spend most of their time inquiring into possible treatment regimens.

108. Rainbolt, G. W. Prescription Drug Laws: Justified Hard Paternalism. *Bioethics,* 1989; 3(1): 55. For a critical discussion of Rainbolt's argument see Ten, C. L. Paternalism and Levels of Knowledge: A Comment on Rainbolt. *Bioethics,* 1989; 3(2): 135-139, as well as Rainbolt's response: 140-141.

109. Arneson, R. Mill versus Paternalism, 1980: 482.

110. Feinberg, J. *The Moral Limits of the Criminal Law: Harm to Self,* 1986: x.

111. Feinberg, J. *Harm to Self,* 1986: 3. Feinberg is exclusively concerned with legal paternalism. Thus, when I use the term paternalism in this paragraph, it always refers to legal paternalism (some sort of state intervention). The discussion concerned with access to unapproved drugs is primarily a debate about legal

paternalism because laws prohibiting the sale, buying, and the use of nonapproved drugs are usually based on paternalistic grounds or public health reasons, which refer to potential harm done to the society by persons who refuse to join clinical trials following established rules of scientific research.

112. Feinberg, J. *Harm to Self*, 1986: 27.

113. Delaney, M. The Case for Patient Access to Experimental Therapy. *Journal of Infectious Diseases*, 1989; 159(3): 416.

114. Feinberg, J. *Harm to Self*, 1986: 28.

115. Feinberg, J. *Harm to Self*, 1986: 30.

116. Feinberg, J. *Harm to Self*, 1986: 33.

117. Feinberg, J. *Harm to Self*, 1986: 32.

118. See Schüklenk, U. Against Manipulative Campaigns by "Community Based" AIDS Organizations. *Health Care Analysis*, 1994; 2(3): 253-261.

119. Feinberg, J. *Harm to Self*, 1986: 32.

120. Feinberg, J. *Harm to Self*, 1986: 39.

121. Feinberg, J. *Harm to Self*, 1986: 39.

122. Feinberg, J. Legal Paternalism, 1971: 105.

123. Feinberg, J. Legal Paternalism, 1971: 106.

124. Arneson, R. Mill versus Paternalism, 1980: 485.

125. All quotes in this and the following three paragraphs have been taken from: Brudner, J. Peptide T Fails in Phase II Test. *Treatment Issues*, 1994; 8(1): 7.

126. Brock, D. W. and Wartman, S. A. When Competent Patients Make Irrational Choices. *New England Journal of Medicine*, 1990; 322(22): 1595-1599 (quote is from 1596).

127. Annas, G. J. Faith (Healing), Hope, and Charity at the FDS: The Politics of AIDS Drug Trials. In Gostin, L. O. (Ed.). *AIDS and the Healthcare System*. New Haven, CT: Yale University Press, 1990, 188.

128. Feinberg, J. *Harm to Self*, 1986: 131. Feinberg defends this view in a similar example of a person who wishes to smoke pot. "The person who has all the relevant knowledge available about the risk of pot smoking does shoulder the risk quite voluntarily, provided also that he has an accurate knowledge, at a higher level, of the scope and limits of his first-level knowledge. . . . Unavoidable ignorance is to some degree an element in all risk-taking, but to know which factors are unknown is itself to have knowledge of a certain kind, contributing to responsible decision-making. All we need to assure ourselves of in assessing voluntariness is that the risk-taker knows exactly what the risk is that he is taking, and his ignorance is a vital component of that risk." 1986: 160-161.

129. Feinberg, J. *Harm to Self*, 1986: 128.

130. Ten, C. L. Paternalism and Levels of Knowledge, 1989: 137.

131. Feinberg, J. *Harm to Self*, 1986: 103. See, for a similar point of view, Buchanan, A. E. and Brock, D. W. *Surrogate Ethics*, 1989: 35 ff.

132. Feinberg, J. *Harm to Self*, 1986: 106.

133. Buchanan, A. E. and Brock, D. W. *Surrogate Ethics*, 1990: 51.

134. Wicclair, M. R. Patient Decision-Making Capacity and Risk. *Bioethics*, 1991; 5(2): 99.

135. Buchanan, A. E. and Brock, D. W. *Surrogate Ethics*, 1990: 34.

136. Anonymous (comment made on-line.). Maury Povich show on ozone therapy. science.medicine.aids poster no. 14334; April 27, 1994, 16:34:02 GMT.

137. Brock, D. W. The Ideal of Shared Decision Making Between Physicians and Patients. *Kennedy Institute of Ethics Journal*, 1991; 1: 32.

138. Garrett, L. Battle on AIDS Drugs. *Newsday*, September 6, 1994: 4 (emphasis added).

139. Tännsjö, T. The Morality of Clinical Research: A Case Study. *The Journal of Medicine and Philosophy*, 1994; 19(1): 12.

140. Tännsjö, T. The Morality of Clinical Research, 1994: 13.

141. Buchanan, A. E. and Brock D. W. *Surrogate Ethics*, 1990: 34.

142. Appelbaum, P. S., Roth, L. H., Lidz, C. W., Benson, P., and Winslade, W. False Hopes and Best Data: Consent to Research and the Therapeutic Misconception. *Hastings Center Report*, 1987; 17(2): 20-4.

143. Appelbaum, P. S., Roth, L. H., Lidz, C. W., Benson, P., and Winslade, W. False Hopes and Best Data, 1987: 20.

144. Kirby, M. Informed Consent: What Does It Mean? *Journal of Medical Ethics*, 1983; 9(2): 73.

145. A randomized, controlled clinical trial is a study in which the consequences of giving one or more experimental agents are analyzed for a given patient group. The agents are allocated to certain patients at random in order to minimize the possibility that factors such as age, severity of illness, race, etc. affect the outcome. Hutton, J. L. The Ethics of Randomized Controlled Trials: A Matter of Statistical Belief? *Health Care Analysis*, 1996; 4(2): 95-102.

146. Wicclair, M. R. Patient Decision-Making and Risk, 1991: 99.

147. Arneson, R. Mill Versus Paternalism, 1980: 475.

148. Dworkin, G. *Theory and Practice*, 1988. I will make use of earlier publications by Dworkin only infrequently because he stresses in his preface to this book the point that the views he has expressed in earlier articles "have undergone considerable revision" (p. xi).

149. Dworkin, G. *Theory and Practice*, 1988: 31.

150. Dworkin, G. *Theory and Practice*, 1988: 20.

151. Frankfurt, H. G. Freedom of the Will and the Concept of a Person. *Journal of Philosophy*, 1971; 68(1): 5-20.

152. Dworkin, G. *Theory and Practice*, 1988: 20.

153. Faden, R. R. and Beauchamp, T. L. *HTIC*, 1986: 263-264.

154. Faden, R. R. and Beauchamp, T. L. *HTIC*, 1986: 266.

155. Dworkin, G. *Theory and Practice*, 1988: 113.

156. Robert Young reaches a similar conclusion in regard to the question of whether Dworkin is supporting only weak paternalistic acts or also strong paternalistic acts. Young, R. *Personal Autonomy*, 1986: 69.

157. Dworkin, G. Paternalism, 1971: 108.

158. Dworkin, G. *Theory and Practice*, 1988: 32, 114.

159. Dworkin, G. *Theory and Practice*, 1988: 117.

160. Dworkin stresses the point that his argument is not based on a hypothetical or assumed consensus of the patient, but on a "set of principles that are justified because they are capable of securing agreement by rational agents under certain circumstances." Dworkin, G. *Theory and Practice*, 1988: 117. This view cannot hide its philosophical background. It is not based in empirical surveys, for instance, that give us an indication how people really act in such circumstances, but on the ideal of the Kantian rational agent.

161. Häyry, H. *The Limits of Medical Paternalism*. London: Routledge, 1991: 183. See, for a similar point of view: Kirby, M. Informed Consent: What Does It Mean? 1983.

162. Beauchamp and Childress warn, with some justification, of possible abuses of the option to waive consent by pointing out that, between patients and physicians, there is generally an atmosphere of trust on the side of the patient that might be utilized on the side of the physician. They write: "A general practice of allowing waivers is dangerous. Many patients have an inordinate trust in physicians, and the general acceptance of waivers of consent in research and therapeutic settings could make patients more vulnerable to those who would abbreviate or omit consent procedures for convenience, already a problem in health care. Accordingly, the danger of abuse of the waiver in busy medical settings, together with problems of how to determine the conditions under which a patient can make a voluntary and informed decision to waive the right to relevant information, demand caution in implementing waivers." Beauchamp, T. L. and Childress, J. F. *Principles*, 1994: 163. This argument might be valid as far as the average patient with a non–life-threatening disease is concerned, but it is certainly not convincing for many people with AIDS who have lost their trust in the integrity of the medical profession entirely. Especially in the case of patients who wish to take experimental agents, it is quite obvious that trust in the work of what is conceived to be the medical establishment has largely ceased to exist. Whether this suspicion is appropriate or prudent is an entirely different question.

163. Archard, D. For Our Own Good. *Australasian Journal of Philosophy*, 1994; 72(3): 292.

164. Archard, D. For Our Own Good, 1994: 292.

165. Feldman, J. The French Are Different—French and American Medicine in the Context of AIDS. *Western Medical Journal*, 1992; 157: 345-349. See also her unpublished manuscript: *An American Anthropologist in a Paris Clinic: Ethical Issues in the French Doctor-Patient Relationship*. University of Illinois, Urbana, Medical Scholar's Program / Dept. of Anthropology, 1992.

166. Hooker, W. *Physician and Patient: Or a Practical View of the Mutual Duties, Relations and Interests of the Medical Profession and the Community*. New York: Baker and Scribner, 1849: 357f. (As quoted in Beauchamp, T. L. and Childress, J. F. *Principles*, 1994: 51.)

167. Dworkin, G. *Theory and Practice*, 1988: 120.

168. Beauchamp, T. L. and Childress, J. F. *Principles*, 1994: 151. They also point out on the same page that "empirical evidence indicates more often than not

that the physician-hypothesized negative effects such as anxiety and reduced compliance do not materialize."

169. Feldman, J. *An American Anthropologist in a Paris Clinic*, 1992: 5.

170. Beauchamp, T. L. and Childress, J. F. *Principles*, 1994: Faden, R. R. and Beauchamp, T. L. *HTIC*, 1986.

171. Beauchamp, T. L. and Childress, James F. *Principles*, 1994: 68.

172. Faden, R. R. and Beauchamp, T. L. *HTIC*, 1986: 238.

173. Faden, R. R. and Beauchamp, T. L. *HTIC*, 1986: 240.

174. Faden, R. R. and Beauchamp, T. L. *HTIC*, 1986: 239.

175. Faden, R. R. and Beauchamp, T. L. *HTIC*, 1986: 243. Dworkin and Young wouldn't object to this definition. Indeed, the idea that an action must follow from such a plan in order to be autonomous is part of their own construction of autonomy. Beauchamp and Childress are right when they claim that a Millian as well as a Kantian approach can "both provide support for the principle of respect for autonomy." *Principles*, 1994: 125.

176. Ten, C. L. Paternalism and Levels of Knowledge, 1989: 136. Ten also believes that prescription drug laws can be given a weak paternalistic foundation. For instance, they could be used in order to find out whether persons lack the necessary first-level knowledge and, if so, provide them with the necessary first-level knowledge. 138-139. Given that his view of the necessity of first-level knowledge for autonomous decision-making fails to convince, so does, inevitably, this conclusion.

177. Häyry, H. *The Limits of Medical Paternalism*, 1991: 180.

178. This can safely be assumed because newsletters of so-called Buyers' Clubs (groups buying and applying illegally unapproved drugs in their attempts to combat AIDS-defining diseases) usually contain firsthand reports of persons who have tried a certain drug in regard to the effects the drug had.

179. Greenberg, J. An Alternative Treatment Activist Manifesto. *Treatment Issues*, 1993; 7(11/12): 1.

180. Greenberg, J. An Alternative Treatment Activist Manifesto, 1993: 1.

181. Armington, K. Evaluating New or Alternative Treatments. *Treatment Issues*, 1993; 7(11/12): 29.

182. Anonymous. Medical Matters: Evaluating Alternative Treatments. People with AIDS Coalition New York City *Newsline*, 1994; March: 26.

183. Wolfstädter, H. D. (Ed.). *Unkonventionelle Medizin bei HIV und AIDS*. Berlin: DAH, 1995. It is worthwhile mentioning that the authors of this book emphasize that they have not accepted any money from any of the producers of the drugs they have discussed. This is, unfortunately, far from common.

184. Hogan, C. The Prudent Pariah: How to Be a Problem Patient. The CBC and diff. *PWALive*, 1994; 5(3): 22. See also Hogan, C. A Professional Patient's Survival Guide. *PWALive*, 1994; 5(3): 19-21.

185. Shorr, A. F. AIDS and the FDA, 1992. Annas, G. Faith (Healing), Hope, and Charity at the FDA, 1990.

186. Anonymous. Medical Matters, 1994: 27.

187. Nussbaum, B. *Good Intentions*, 1990: 231-235.

188. The state of the art today is undoubtedly a ranking of Bactrim (TMP/SMX), dapsone, and last, aerosolized pentamidine.
189. Carter, G. M. Stinkhorn. science.medicine.aids poster no. 11833, April 18, 1995.
190. Kwitny, J. *Acceptable Risks*. New York: Poseidon Press, 1992: 19.
191. Hodel, D. The Dietary Supplement Health and Education Act of 1993 (S.784/H.R. 1709). *Treatment Issues*, 1993; 7(11/12): 31.
192. Cox, S. Personal communication. February 24, 1994.
193. I will get back to the problem of whether, therefore, a risk-related standard of competence is the appropriate answer to this situation later in this chapter.
194. Dobson, P. Personal communication. February 25, 1994.
195. As I have shown elsewhere, self-proclaimed "community based" AIDS organizations, and "community leaders" indeed employ unjustifiably manipulative techniques in their various AIDS educational campaigns in order to ensure that persons with AIDS adhere to these organizations' rules of conduct. See Schüklenk, U. Against Manipulative Campaigns by "Community Based" AIDS Organizations, 1994: 253-261.
196. Beauchamp, T. L. and Childress, James F. *Principles*, 1994: 214, 216.
197. Beauchamp, T. L. and Childress, James F. *Principles*, 1994: 220.
198. See also the similar situation discussed earlier in the paragraph devoted to the Youngian theory.
199. Young, R. *Personal Autonomy*, 1986: 21.
200. Young, R. *Personal Autonomy*, 1986: 109 (emphasis added).
201. Young, R. *Personal Autonomy*, 1986: 31 (emphasis added).
202. Young, R. *Personal Autonomy*, 1986: 34. See, as mentioned earlier, the overall convincing critical response of Faden, R. R. and Beauchamp, T. L. *HTIC*, 1986: 262-269.
203. Young, R. *Personal Autonomy*, 1986: 7 (see here also the references to the relevant works of Joel Feinberg and Harry Frankfurt).
204. National Research Council. *Social Impact of AIDS*, 1993.
205. Young, R. *Personal Autonomy*, 1986: 8.
206. Young interprets the expression life plan as follows: "The term 'plan' is here intended to refer merely to whatever it is that a person broadly wants to do in his or her life—thus covering career, lifestyle, dominant pursuits and the like, not some inflexible or unchanging set of goals." See his *Personal Autonomy*, 1986: 8. For a similar notion of autonomy see Benn, S. I. Freedom, Autonomy, and the Concept of a Person. *Proceedings of the Aristotelian Society*, 1976; LXXVI: 123f. He wrote: "The autonomous man is the one . . . whose life has a consistency that derives from a coherent set of beliefs, values, and principles, by which his actions are governed. . . . The principles by which the autonomous man governs his life make his decisions consistent and intelligible to him as his own; for they constitute the personality he recognizes as the one he has made his own" (p. 123f).
207. Young, R. *Personal Autonomy*, 1986: 65.
208. John Kleinig presents us with a nearly identical defense of paternalistic interventions: "Where our conduct or choices place our permanent, stable, and

central projects in jeopardy, and where what comes to expression in this conduct or these choices manifests aspects of our personality that do not rank highly in our constellation of desires, dispositions etc., benevolent interference will constitute no violation of integrity. . . . It helps to preserve it." Kleinig, J. *Paternalism*, 1984: 68. For a critical response see Rainbolt, G. W. Prescription Drug Laws: Justified Hard Paternalism, 1989: 48. See also Archard, D. For Our Own Good, 1994: 290.

209. Young, R. *Personal Autonomy*, 1986: 72.
210. Young, R. *Personal Autonomy*, 1986: 74.
211. Young, R. *Personal Autonomy*, 1986: 76.
212. Young, R. *Personal Autonomy*, 1986: 75.
213. Feinberg, J. Legal Paternalism, 1971: 120.
214. Young, R. *Personal Autonomy*, 1986: 78.
215. Dixon, J. Catastrophic Rights: Vital Public Interests and Civil Liberties in Conflict. Overall, C. and Zion, W. P. (Eds.). *Perspectives on AIDS: Ethical and Social Issues*. Toronto: Oxford University Press, 1991: 130f. Dixon argues basically that, before we discuss giving patients access to experimental agents, we should try to accommodate their interests better than we do currently. He believes that this might help to avoid the known cheating by patients in clinical trials and it might eliminate, by and large, the interest of terminally ill patients to access experimental agents outside clinical trials: 136.

216. Young, R. *Personal Autonomy*, 1986: 78.
217. Young, R. *Personal Autonomy*, 1986: 79.
218. He served as president of the Voluntary Euthanasia Society in the Australian state of Victoria. See also a paper co-authored with Helga Kuhse: Kuhse, H. and Young, R. Options for Dying with Dignity. *Health Issues,* September 1987: 8-9.

219. It is true that this may be restricted to the problem of buying and applying experimental drugs and that persons with AIDS, apart from that, are free to live their lives as they wish to. However, practically, it would be necessary to monitor persons with AIDS on a twenty-four-hour basis in order to enforce the restriction imposed upon them. Additionally, we would restrict their dispositional autonomy not in a minor area (such as preventing them from seeing the latest Arnold Schwarzenegger movie), but in an area that is of central importance to them, which is the treatment of the diseases which are likely to kill them in the near future. Hence, to them, it is the question of life or death.

220. Of course, average implies that some exceptional people might survive much longer. Reports are around of so-called long-term survivors who are said to have survived AIDS longer than a decade. However, these are exceptions that are counterbalanced by the overwhelming number of patients who die very quickly once they have developed full-blown AIDS.

221. As I will demonstrate at great length in the next chapter, this standard claim, made in support of some currently available and used standard AIDS treatments, is more than questionable.

222. Darby, A., Blake, E. and Dow, S. AMA Push for States to Reject Euthanasia. *The Age,* May 26, 1995: A6.

223. The only exception is one medical condition, AIDS dementia, which renders some AIDS patients incompetent.

224. The only exception that I have mentioned is the case of a patient who has a mistaken idea of what a certain experimental agent might do. In other words, if there is a drug that has a very high likelihood of killing or significantly harming a person with a severely compromised immune system, and there is no evidence that this drug has any positive effects on a given patient, and this patient argues that it will save his life, then clearly there is a reason to intervene on the basis of the arguments presented by Mill in the famous bridge example that I discussed earlier in this chapter.

225. I consider knowledge sufficient when the patient who wishes to try an experimental drug is aware of the fact that finally it remains unclear what impact this drug will have on his or her health status.

226. It is a necessary condition that criterion (1) is met.

227. Young, R. *Personal Autonomy*, 1986: 47.

Chapter 2

1. Cooper, E. Controlled Clinical Trials of AIDS Drugs: The Best Hope. *JAMA*, 1989; 261: 2445.

2. Dixon, J. Catastrophic Rights: Vital Public Interests and Civil Liberties in Conflict. In Overall, C, and Zion, W. P. (Eds.). *Perspectives on AIDS: Ethical and Social Issues*. Toronto: Oxford University Press, 1991: 131.

3. Dixon, J. *Catastrophic Rights*, 1991: 132.

4. For a fully extended presentation of this persuasive interpretation of AIDS acquisition being mostly a harm to self situation, see Mohr, R. D. AIDS, Gays, and State Coercion, *Bioethics*, 1987; 1(1): 35-50; and Illingworth, P. *AIDS and the Good Society*, London: Routledge, 1990.

5. Annas, G. Faith (Healing), Hope, and Charity at the FDA: The Politics of AIDS Drug Trials. In Gostin, L. O. (Ed.). *AIDS and the Healthcare System*. New Haven: Yale University Press, 1990: 190.

6. See, for example: Arras, J. D. Noncompliance in AIDS Research. *Hastings Center Report*, 1990; 20(5): 24-32; Delaney, M. The Case for Patient Access to Experimental Therapy. *Journal of Infectious Diseases*, 1989; 159: 416-419.

7. Delaney, M. The Case for Patient Access, 1989: 418. I will discuss this ethical problem in the paragraph on coercive offers, which occurs at the end of this chapter. Recently, researchers have come to acknowledge the phenomenon described by Delaney. They argue that a master strategy for clinical trials "must take account of the inevitable fact that patients and clinicians may decide to change from the study therapy to which the patient was randomized. It is not the purpose of this structure to encourage such changes: there is no evidence that change for its own sake benefits patients, and changes tend to dilute treatment differences and necessitate large sample sizes. However, *we must recognize that some patients and clinicians will feel compelled to change therapy. Thus, it is important to accommodate the effect of changes in therapy and, whenever pos-*

sible to address questions about whether change is beneficial." Hogan, C. and Hodges, J. CPCRA Master Antiretroviral Protocol Strategy (MAPS), Draft #4. Unpublished manuscript, Minneapolis: University of Minnesota, Division of Biostatistics, School of Public Health, September 23, 1993: 7 (italics added).

8. Novick, A. Reflections on a Term of Public Service with the FDA Anti-Virals Advisory Committee. *AIDS and Public Policy Journal,* 1993; 8(2): 58-59.

9. Dixon, J. *Catastrophic Rights,* 1991: 123.

10. Arras, J. D. Noncompliance in AIDS Research, 1990: 29. (Emphasis added.)

11. Delaney, M. The Case for Patient Access to Experimental Therapy, 1989: 417. That many people with AIDS hold this opinion is also supported in an article by David A. Salisbury and colleagues, who maintain that "the randomized format is thought to be coercive [by patients] because subjects are forced into the trial as the only way to receive the experimental agent." Salisbury, D. A. and Schechter, M. T. AIDS Trials, Civil Liberties and the Social Control of Therapy: Should We Embrace New Drugs with Open Arms? *Canadian Medical Association Journal,* 1990; 142(10): 1059.

12. 1993 Revised Classification System for HIV Infection and Expanded Surveillance Case Definition for AIDS Among Adolescents and Adults. *Morbidity and Mortality Weekly Report,* 1992; 41(RR-17): 1-19.

13. Lenderking W. R., Gelber R. D., Cotton D. J., Cole, B. F., Goldhirsch, A., Volberding, P. A., and Testa, M. A. Evaluation of the Quality of Life Associated with Zidovudine Treatment in Asymptomatic Human Immunodeficiency Virus Infection. *New England Journal of Medicine,* 1994; 330(11): 738-743.

14. Abrams, D. I. On the Matter of Survival. *Beta: Bulletin of Experimental Treatments,* 1992; November: 12.

15. Hellman, S. and Hellman D. S. Of Mice but Not Men: Problems of the Randomized Clinical Trial. *New England Journal of Medicine,* 1991; 324: 1585-1589.

16. Hellman, S. The Patient and the Public Good. *Nature Medicine,* 1995; 1(5): 402.

17. Gray, J. N., Lyons, P. M., Jr., and Melton, G. B. *Ethical and Legal Issues in AIDS Research.* Baltimore, MD: The Johns Hopkins University Press, 1995: 35.

18. Mirken, B. AIDS Clinical Trials: Why They Have Recruiting Problems. *AIDS Treatment News,* 1995; 217: 1-4, quote on p. 4.

19. This, of course, is impossible, even if I attempt to disable their legally possible access to experimental drugs by making it illegal, as it is in most countries today. It seems unreasonable to believe that we could stop a desperate, terminally ill person from breaking the law if this person thinks this is the only chance to save his or her life.

20. Shorr, A. F. AIDS and the FDA: The Ethical Case for Limiting Patient Access to New Medical Therapies. *IRB: A Review of Human Subjects Research,* 1992; 14(4): 2.

21. Annas, G. J. Faith (Healing), Hope, and Charity, 1990: 184.

22. Appelbaum, P. S., Roth, L. H., Lidz, C. W., Benson, P., and Winslade, W. False Hopes and Best Data: Consent to Research and the Therapeutic Misconception. *Hastings Center Report,* 1987; 17(2): 20.

23. Appelbaum, P. S, Roth, L. H., Lidz, C. W., Benson, P., and Winslade, W. False Hopes and Best Data, 1987: 20.

24. Novick, A. Reflections, 1993: 58.

25. Jonas, H. Philosophical Reflections on Experimenting with Human Subjects. *Daedalus,* 1969; 98: 219-247.

26. Schafer, A. Experimenting with Human Subjects: A Critique of the Views of Hans Jonas. *Journal of Medical Ethics,* 1983; 9:76-79.

27. For a full philosophical exploration of a utilitarian view of all facets of this problem, see: Birnbacher, D. *Verantwortung für zukünftige Generationen.* Stuttgart, Germany: Reclam Verlag, 1988.

28. This is true at least under normal circumstances. A few famous exceptions, such as Mahatma Gandhi and some of his disciples, are not good proof (and definitely not a persuasive argument) for the claim that such noble behavior is also possible for the average citizen today who is suffering a terminal disease.

29. Singer, P. *Practical Ethics,* Second Edition. Cambridge, MA: Cambridge University Press, 1993: 229.

30. Arras, J. D. Noncompliance in AIDS research, 1990: 29.

31. Singer, P. *Practical Ethics,* 1993: 246.

32. Foot, P. The Problem of Abortion and the Doctrine of the Double Effect. Steinbock, B. (Ed.). *Killing and Letting Die.* Englewood Cliffs, NJ: Prentice-Hall, 1980: 159.

33. Rawls, J. *A Theory of Justice.* Cambridge, MA: Harvard University Press, 1971.

34. Tännsjö, T. The Morality of Clinical Research: A Case Study. *Journal of Medicine and Philosophy,* 1994; 19(1): 19.

35. Hellman, S. and Hellman, D. S. Of Mice but Not Men, 1991: 1856.

36. One may also note that the deliberations took place exclusively between clinicians and a moral philosopher. Seemingly the patients were not deemed appropriate partners for a debate about the acceptability of a trial arm that offered a 1/100 to 1/1000 chance of death from the administered drug.

37. Hellman, S. The Patient and the Public Good. *Nature Medicine,* 1995: 400.

38. My view on this matter is similar to that of Samuel Hellman, who suggested that we should strictly separate the role of the principal investigator from that of the primary care physician. He argues that "these two roles residing in the same physician create conflicting goals, which have the potential to undermine the primacy of the patient." Hellman, S. The patient and the public good, 1995: 402.

39. When I deny these patients access to experimental drugs, I rob them of this option to decide. They are left with the choice of either joining a badly designed clinical trial or dying without any access to drugs that may offer a slim chance of success.

40. Dixon, J. *Catastrophic Rights*, 1991: 135.

41. Nussbaum describes at some length appallingly designed clinical AIDS trials; the principal investigators of which had trouble in recruiting the necessary number of research subjects. See: Nussbaum, B. *Good Intentions: How Big Business and the Medical Establishment Are Corrupting the Fight Against AIDS, Alzheimer's, Cancer, and More.* New York: Penguin, 1990. Similar information can be found in: Anonymous. Zu wenig Probanden für AZT-Versuch. *Die Neue Ärztliche*, January 4, 1988. Unfortunately, it seems, the situation has not improved significantly over the last decade. A more recent article in *AIDS Treatment News* reports: "This [not enough enrolled patients] is a significant problem. Collection of needed data can be greatly delayed. Worse, if trials are underenrolled, or if the factors that cause people not to enroll also lead large numbers of volunteers to drop out or cheat, either by lying to get into the study or by not following the trial regimen, the result could be data that is of little or no value. Obstacles to enrollment in AIDS trials generally fall into one of two broad areas: The publicity or outreach efforts used to recruit prospective volunteers, and the design of the trials themselves." Mirken, B. AIDS Clinical Trials, 1995: 1.

42. This is in contrast to diseases of a more harmless nature, where the possible negative consequences of a certain experimental therapy are reversible.

43. Quoted in: Cameron, M. E. *Living with AIDS—Experiencing Ethical Problems.* Newbury Park, CA: Sage, 1993: 117-118. Most other persons living with AIDS-defining diseases expressed similar views. The general attitude was that they would not trust their physician or principal investigator because they are "trying to take our money," as "Nancy" said or, in the view of "Bill," those scientists do not tell the truth "because of money."

44. HIV-Support e-mailing list, poster sent on August 10, 1995. This same patient concluded his poster after reporting his experiences in another trial saying that he would not participate ever again in a clinical trial. "No more I say. Life is too precious to do this. I can honestly say that I am sorry that I did this."

45. Novick, A. Reflections, 1993: 58-59.

46. Rothman, K. J. and Michels, K. B. The Continuing Unethical Use of Placebo Controls. *New England Journal of Medicine*, 1994; 331(6): 394.

47. Mirken, B. AIDS Clinical Trials, 1995: 3-4.

48. As quoted in Mirken, B. AIDS Clinical Trials, 1995: 3-4.

49. See ACT UP's argument that every clinical trial, at least in the case of AIDS or any other terminal disease, is always also to be seen as a therapy and never only as a research enterprise. George J. Annas summarized this point of view quite correctly: "The position of the gay community is probably best summed up in a slogan used by AIDS Coalition to Unleash Power (ACT UP): A Drug Trial Is Health Care Too." Annas, G. J. Faith (Healing), Hope, and Charity, 1990: 186. Felicia R. Lee correctly pointed out in *The New York Times* that "what scientists see as research, AIDS patients see as survival" (B7). Lee, F. R. U.S. Cuts Three of Four AIDS Research Grants for Poor New Yorkers. *The New York Times*, November 8, 1994: A1, B7. It seems to be the case that, under the current health care system in the United States, poor patients are the first to take their

chances in research clinical trials because they cannot afford experimental therapies. Thus, even if experimental therapies could freely be purchased in this country, they would still have no other choice but to join clinical trials for that simple reason. In my subsequent discussion about who should pay for access to experimental therapies, I will return to this problem of social inequalities in countries without universal health care.

50. Victorian AIDS Council/Gay Men's Health Centre. Not All Clinical Trials Go Horribly Wrong. *Melbourne Star Observer,* 1995; 251 (suppl. Positive Living): 8.

51. Lee, F. R. U.S. Cuts Three of Four AIDS Research Grants, 1994: B7.

52. This, however, should not be confused with *any* poor patient with AIDS. A study undertaken by Yale University's Jeanett Ickovics concluded that "Latinos, and also current or injecting drug users, were least likely to have been asked to participate." Ickovic's study reports furthermore, however, that common stereotypes about injecting drug users allegedly being unreliable are false. She writes "Neither sex, race nor history of injecting drug use were associated with nonadherence to medication regimen." Nevertheless, "Injecting drug users are often excluded from clinical trials by exclusion criteria that explicitly (e.g. no active drug users) or implicitly (e.g. high standards for liver function test) eliminate them from clinical trials." In particular, in cases of studies of remedies that are deemed promising in the AIDS communities, usually no attempt is made by researchers to reach a representative cross-section of patients. Jeff Getty, a treatment activist in San Francisco who volunteers at the Oakland AIDS Services Center believes that there is an "unspoken prejudice that's quietly there, keeping women, poor and IDUs out of these studies." This is pretty easy to achieve by simply not accommodating these patients' needs for reimbursement for travels across the bay to San Francisco, or not giving allowances for child care while the mother is hospitalized, etc. As I will show in the subsequent paragraphs, this selection of patients leads inevitably to the problem that we simply cannot know, from such a study, how a certain agent tested works in populations such as ethnic minorities, women, IDUs, and so on and so forth. All quotes are from: Mirken, B. AIDS Clinical Trials, 1995: 3.

53. Fischl, M. A., Richman, D. D., Grieco, M. H., Gottlieb, M. S., Volberding, P. A., Laskin, O. L., Leedom, J. M., Groopman, J. E., Mildvan, D., Schooley, R. T. et al. The Efficacy of Azidothymidine (AZT) in the Treatment of Patients with AIDS and AIDS-Related Complex. *New England Journal of Medicine,* 1987; 317(4): 185-191. Richman, D. D., Fischl, M. A., Grieco, M. H., Gottlieb, M. S., Volberding, P. A., Laskin, O. L., Leedom, J. M., Groopman, J. E., Mildvan, D., Hirsch, M. S. et al. The Toxicity of Azidothymidine (AZT) in the Treatment of Patients with AIDS and AIDS-Related Complex. *New England Journal of Medicine,* 1987; 314(4): 192-197.

54. This is commonly known under the names azidothymidine, zidovudine, 3'-azido3'-deoxythimidine, and the brand name, Retrovir.

55. All patients in this trial would today be defined as AIDS patients or persons with AIDS. The technical term ARC has been replaced by incorporating

its criteria in the AIDS definition, which now contains three clinical categories in which patients are placed, mainly for surveillance purposes. Compare Centers for Disease Control. 1991 Revised Classification System, 1992: RR-17, 3-4.

56. Sonnabend, J. A. Review of AZT Multicenter Trial—Data Obtained Under the Freedom of Information Act by Project Inform and ACT UP. *AIDS Forum,* 1989; 1(1): 9.

57. Badewitz-Dodd, L. H. (Ed.). *MIMS Annual 1994.* Australian edition. Crows Nest, Australia: MediMedia, 1994. Part 8.I Antiviral agents: 520-521.

58. Arky, R. et al. (Eds.). *Physicians Desk Reference,* Forty-Eighth Edition. Montvalle: Medical Economics Data Production Company, 1994.

59. Bruce Nussbaum argued quite persuasively that about 46 percent of the patients who were on AZT required blood transfusions. Most of them needed more than one transfusion. The problem is to distinguish between rises in CD4+ counts that were due to the alleged beneficial effect of AZT and those that were a consequence of the blood transfusions. After all, the transfusion blood contained CD4+ cells, and the Phase II trial did not control for this. Nussbaum, B. *Good Intentions,* 1990.

60. Fischl, M. A., Richman, D. D., Grieco, M. H., Gottlieb, M. S., Volberding, P. A., Laskin, O. L., Leedom, J. M., Groopman, J. E., Mildvan, D., Schooley, R. T. et al. The Efficacy of Azidothymidine in the Treatment of Patients with AIDS and AIDS Related Complex, 1987: 186.

61. Nussbaum, B. *Good Intentions,* 1990: 123.

62. Osmond, D., Charlebois, E., Lang, W., Shiboski, S., and Moss, A. Changes in AIDS Survival Time in Two San Francisco Cohorts of Homosexual Men, 1983 to 1993. *JAMA,* 1994; 271(14): 1083.

63. Nussbaum, B. *Good Intentions,* 1990: 170.

64. Anonymous. Zidovudine Again. *Lancet,* 1992; 339: 1105-1106.

65. Lauritsen, J. *The AIDS War,* Second Edition. New York: Asklepios Press, 1993: 383.

66. Lauritsen, J. *The AIDS War,* 1993: 77-78.

67. Lauritsen, J. *The AIDS War,* 1993: 78.

68. As reported by: Lauritsen, J. *The AIDS War,* 1993: 74.

69. Fischl, M. A., Richman, D. D., Grieco, M. H., Gottlieb, M. S., Volberding, P. A., Laskin, O. L., Leedom, J. M., Groopman, J. E., Mildvan, D., Schooley, R. T. et al. The Efficacy of Azidothymidine, 1987 185. (See, for example, the subheadline: "A Double-Blind, Placebo-Controlled Trial.") Richman, D. D., Fischl, M. A., Grieco, M. H., Gottlieb, M. S., Volberding, P. A., Laskin, O. L., Leedom, J. M., Groopman, J. E., Mildvan, D., Hirsch, M. S. et al. The Toxicity of Azidothymidine, 1987: 192. (See, for example, the subheadline: "A Double-Blind, Placebo-Controlled Trial.") That these trials became (mistakenly) widely accepted in the biomedical communities as double-blind and placebo-controlled is demonstrated by the aforementioned quote by Anthony Fauci just as much as by recently published review articles. See, for example, McLeod, G. X., Hammer, S. M. Zidovudine: Five Years Later. *Annals of Internal Medicine,* 1992; 117: 487-501; or Sande, M. A., Carpenter, C. C., Cobbs, G. C., Holmes, K. K., and

Sanford, J. P. Antiretroviral Therapy for Adult HIV-Infected Patients: Recommendations from a State of the Art Conference. *JAMA*, 1993; 270(21): 2583-1589.

70. I am not suggesting that a placebo is per se equivalent to nothing. Placebos can have all sorts of fascinating positive effects, as well as devastating negative effects. The reasons for this phenomenon are not fully understood as of yet. However, a placebo is equal to nothing when patients (and sometimes physicians) know who is using the active experimental agents and who has been randomized to the placebo arm. This was obviously the case in the study I am considering here. The so-called placebo effect could not have come into operation because patients knew who received a placebo and who got the active experimental agent. It is therefore reasonable to argue that AZT was compared with nothing.

71. Sonnabend, J. A. Review of AZT Multicenter Trial, 1989: 12.

72. As quoted in: Lauritsen, J. *The AIDS War*, 1993: 389.

73. Sonnabend, J. A. Review of AZT Multicenter Trial, 1989: 10.

74. Sonnabend, J. A. Review of AZT Multicenter Trial, 1989: 12.

75. Lauritsen, J. *The AIDS War*, 1993: 389.

76. Sonnabend, J. A. Review of AZT Multicenter Trial, 1989: 13-14.

77. Burroughs-Wellcome has undergone a merger with the pharmaceutical company Glaxo and is now known under the name Glaxo-Wellcome. Whenever I make mention of Wellcome or Burroughs-Wellcome in this book, I refer to the company before the merger.

78. Arky, R. et al. (Eds.). *Physicians Desk Reference*, 1994: 745.

79. Okie, S. Higher Cancer Rate Seen in Users of AIDS Drug. *The Washington Post*, August 16, 1990.

80. Yarchoan, R., Hiroaki, M., Myers, C. E., Broder, S. Clinical Pharmacology of 3',-Azido-2',3'-Dideoxythymidine (Zidovudine) and related Dideoxynucleosides. *New England Journal of Medicine*, 1989; 321: 731.

81. Pluda, J. M., Yarchoan, R., Jaffe E. S., Feuerstein I. M., Solomon, D., Steinberg, S., Wyvill, K. M., Raubitschek, A., Katz, D., and Broder, S. Development of Non-Hodgkin Lymphoma in a Cohort of Patients with Severe Human Immunodeficiency Virus Infection on Long-Term Antiretroviral Therapy. *Annals of Internal Medicine*, 1990; 113: 276-282, quote is from p. 278.

82. Duesberg, P. H. AIDS: Acquired by Drug Consumption and Other Non-contagious Risk Factors. *Pharmacology and Therapeutics*, 1992; 55: 244.

83. Okie, S. Higher Cancer Rate, 1990.

84. Centers for Disease Control. Opportunistic Non-Hodgkin's Lymphomas Among Severely Immunocompromised HIV-Infected Patients Surviving for Prolonged Periods on Antiretroviral Therapy—United States. *Morbidity and Mortality Weekly Report*, 1991; 40(34): 558-559.

85. Duesberg, P. H. AIDS, 1992: 244.

86. Cote, T. R. and Biggar, R. J. Does Zidovudine Cause Non-Hodgkin's Lymphoma? *AIDS*, 1995; 9(4): 404-405.

87. National Cancer Institute. RFA: CA-95-009. P.T. 34; K.W. 075008, 0715035, 0755015. *NIH Guide*, 1995; 24(7): no pages numbers available.

88. Abrams, D. I. On the Matter of Survival, 1992: 12. I asked Professor Abrams to define more clearly what the difference between mortality and survival is. He replied, "Cumulative mortality is one minus cumulative survival. If the one year cumulative mortality is 0.30 (30 percent), then the one year cumulative survival is 70 percent. When we speak of rates, we mean per unit time, for example, the death rate per one hundred person years. There is no corresponding survival rate. For example, if the rate is ten per one hundred person years—sometimes written as 10.0 in tables—then 90.0 is not the survival rate. My personal emphasis on survival is as an endpoint in trials as compared to surrogate endpoints such as CD4+ cell counts, etc." Personal communication, March 7, 1995.

89. Anonymous. Zidovudine Again, 1992: 1106.

90. Hamburg, M. A. and Fauci, A. S. HIV Infection and AIDS: Challenges to Biomedical Research. In Gostin, L. O. (Ed.). *AIDS and the Healthcare System*. New Haven: Yale University Press 1990: 178.

91. Abrams, D. I. Survival: The Ultimate Endpoint. *AIDS/HIV Treatment Directory*, 1993; 6(2): 5.

92. Choi, S., Lagakos, S. W., Schooley, R. T., and Volberding, P. A. CD4+ Lymphocytes Are an Incomplete Surrogate Marker for Clinical Progression in Persons with Asymptomatic HIV Infection Taking Zidovudine. *Annals of Internal Medicine*, 1993; 118(9): 674.

93. Concorde Coordinating Committee. Concorde: MRC/ANRS Randomized Double-Blind Controlled Trial of Immediate and Deferred Zidovudine in Symptom-Free HIV Infection. *Lancet*, 1994; 343: 871.

94. Concorde Coordinating Committee. Concorde, 1994: 871, italics added.

95. Anonymous. Protein Seems to Stimulate AIDS-Damaged Immune-Systems. *Reuter Newsagency Dispatch*, March 2, 1995 (2300 GMT).

96. Kovacs, J. A., Dewar, R. J., Vogel, S., Davey, R. T., Falloon, J., Polis, M. A., Walker, R. E., Stevens, R., Salzman, N. P. et al. Increases in CD4+ Lymphocytes with Intermittent Courses of Interleukin-2 in Patients with Human Immunodeficiency Virus Infection. *New England Journal of Medicine*, 1995; 332: 567.

97. Abrams, D. I. Survival: The Ultimate Endpoint, 1993: 6.

98. Gonsalves, G. Viral Assay Reimbursements. science.medicine.aids article no. 11755, posted April 7, 1995.

99. Neaton, J. D., Wentworth, D. N., Rhame, F., Hogan, C., Abrams, D. I., and Deyton, L. Considerations in Choice of a Clinical Endpoint for AIDS Clinical Trials. *Statistics in Medicine*, 1994; 13(19-20): 2107-2125.

100. Arky, R. et al. (Eds.). *Physicians Desk Reference*, 1994: 744.

101. Not surprisingly, bioethicists have also been misled by these advertisements. James F. Childress, Ronald Bayer, and Carol Levine, for example, have reevaluated test campaigns and come to much more positive conclusions in regard to HIV testing, simply because "early intervention in HIV disease" seemed available. See, for example, Levine, C. and Bayer, R. The Ethics of Screening for Early Intervention in HIV Disease. *American Journal of Public Health*, 1989; 79: 1601-1607; and Childress, J. F. Mandatory HIV Screening and Testing. Beau-

champ, T. L. and Walters, L. (Eds.). *Contemporary Issues in Bioethics*, Fourth Edition. Belmont, CA: Wadsworth Publ., 1994: 557-571.

102. Anonymous. British Widow Sues U. S. Agency Over AZT. *The Reuters European Business Report*, January 28, 1994.

103. Havlir, D. and Richman, D. D. Antiretroviral Therapy. *Current Opinion in Infectious Diseases*, 1995; 8: 66-73.

104. Volberding, P. A. Is It Possible to Prove a Survival Benefit from Early Treatment? *Bulletin of Experimental Treatment for AIDS*, 1992; November: 15.

105. Even under such circumstances, however, it seems doubtful that patients will be prepared to accept a situation where a rigorous scientific evaluation is made "to the detriment of HIV-positive persons and their medical surveillance." TRT-5. Concerning a Clinical Phase III Trial with Saquinavir: Balancing Research and Health Care. science.medicine.aids article no. 11418. Forwarded by Mark Buzza. Posted February 22, 1995.

106. It may well be argued (and be true) that, under such circumstances, it would be impossible to find these truly altruistically acting volunteers in the first place. In this case, I would argue, we must assume that the members of our societies are unwilling to act altruistically at significant costs for their own lives. Should this be true, I think, it does also indicate that there is not even a societal consensus supporting the fact that we already force an unfortunate few into trials that nobody would want to join were alternatives available. Hence, I cannot think of an ethical justification for current legislations that prevent patients with terminal illnesses from access to experimental agents in order to force them by default into the participation in research clinical trials.

107. Anonymous. British Widow Sues U. S. Agency Over AZT, 1994.

108. Perhaps I should make clear at this point that I do not claim that all the open questions and doubts which I have asked and raised in regard to this study have never before been addressed. Some of the issues mentioned earlier have been satisfactorily answered, while others have not. The points that I wish to make are that they should have been answered before the approval of AZT and that, ultimately, the problematic standards accepted by the FDA for this drug have led to even more drugs being approved without proven efficacy or known toxicity.

109. Brook, I. Approval of Zidovudine (AZT) for Acquired Immunodeficiency Syndrome: A Challenge to the Medical and Pharmaceutical Communities. *JAMA*, 1987; 258(11): 1517. (Emphasis added.)

110. Nussbaum, B. *Good Intentions*, 1990: 170.

111. Editorial. On Stopping a Trial Before Its Time. *Lancet*, 1993; 342: 1311.

112. Hodel, D. FDA Reform Floated in DC. *Treatment Issues*, 1995; 9(6): 6.

113. Current suggestions to overcome this problem include the suggestion to shift efficacy studies to a postmarketing setting whereby at least sufficient information about the toxicities of an experimental agent are necessary before the FDA can give approval. Of course, this would not prevent people with AIDS from accessing unapproved drugs illegally, but it is a step in the right direction. See, for a discussion of this and other proposals, Hodel, D. FDA Reform Floated in DC, 1995.

114. Editorial. On Stopping a Trial Before Its Time, 1993: 1311.

115. Hamilton, J. D., Hartigan, P. M., Simberkoff, M. S., Day, P. L., Diamond, G. R., Dickinson, G. M., Drusano, G. L., Egorin, M. J., George, W. L., Gordin, F. M. et al. A Controlled Trial of Early Versus Late Treatment with Zidovudine in Symptomatic Human Immunodeficiency Virus Infection. *New England Journal of Medicine,* 1992; 326: 437-443. This study ran over a period of four years. Concorde Coordinating Committee. Concorde: MRC/ANRS Randomized Double-Blind Controlled Trial, 1994. This study ran from October 1988 until October 1991, with a participation of 1,749 HIV-positive individuals. Followup was to death or December 1992.

116. Egger, M., Neaton, J. D., Phillips, A. N., and Smith. G. D.. Concorde Trial of Immediate versus Deferred Zidovudine. *The Lancet,* 1994; 343: 1355.

117. D'Aquila, R. T., Johnson, V. A., Welles, S. L., Japour, A. J., Kuritzkes, D. R., DeGruttola, V., Reichelderfer, P. S., Coombs, R. W., Crumpacker, C. S., Kahn, J. O. et al. Zidovudine Resistance and HIV-1 Disease Progression During Antiretroviral Therapy. *Annals of Internal Medicine,* 1995; 122(6): 401-408.

118. Hogan, C. Test Treatment Strategies under Real Conditions. *Treatment Issues,* 1995; 9(4): 7.

119. Pence, G. E. *Classical Cases in Medical Ethics: Accounts of Cases That Have Shaped Medical Ethics, with Philosophical, Legal, and Historical Background,* Second Edition. New York: McGraw-Hill, 1995: 434. Pence refers here in particular to this study: Lenderking, W. HIV-Infection and Long-Term Morbidity. *New England Journal of Medicine,* 1994; March 18, and to an article by Altman, L. Experts Change Guidelines to Using Drugs for HIV. *The New York Times,* June 27, 1993: A1.

120. TRT-5. Concerning a Clinical Phase III Trial of Saquinavir, 1995.

121. TRT-5. Concerning a Clinical Phase III Trial of Saquinavir, 1995.

122. Abrams, D. I. On the Matter of Survival, 1992: 13.

123. Fleming, T. R. Surrogate Markers in AIDS and Cancer Trials. *Statistics in Medicine,* 1994; 13: 1425.

124. DeGruttola, V., Wolfsohn, M., Fischl, M. and Tsiatis, A. Modeling the Relationship Between Survival and CD4+-Lymphocytes in Patients with AIDS and ARC. *Journal of AIDS,* 1993; 6: 359-365.

125. Fischl, M. A., Richman, D. D., Grieco, M. H., Gottlieb, M. S., Volberding, P. A., Laskin, O. L., Leedom, J. M., Groopman, J. C., Mildvan, D., Schooley, R. T. et al. The Efficacy of Azidothymidine in the Treatment of Patients with AIDS and AIDS Related Complex, 1987: 186.

126. Sonnabend, J. A. Do We Need New Ways to Evaluate Experimental AIDS Treatments? *AIDS Forum,* 1989; 1(1): 4-5.

127. See: Appelbaum, P. S., Roth, L. H., Lidz, C. W., Benson, P., and Winslade, W. False Hopes and Best Data, 1987.

128. Lenderking W. R., Gelber R. D., Cotton D. J., Cole, B. F., Goldhirsch, A., Volberding, P. A., and Testa, M. A. Evaluation of the Quality of Life Associated with Zidovudine Treatment, 1994: 738.

129. Baenen, L. Off AZT. *Associated Press,* May 16, 1994.

130. Blazey, P. AZT: A Cure for Dandruff? *With Complements*, 1994; 3(2): 2.
131. Brown, S. Study says Wellcome's AZT Ineffective. *Reuters Financial Report*, January 30, 1994.
132. I assume that individuals make very different decisions in regard to the question of which (if any) losses of quality of life they are willing to trade in order to live longer. Hence, it is useful to gather information as detailed as possible to allow patients to make such decisions. I do not see how the second sentence is a corollary of the first (as indicated by the "hence")—why should the fact of different decisions lead us to conclude that we need to provide particularly detailed info?
133. I should also point out that these are the most important questions that other trials investigating AIDS drugs must also answer. AZT is the example that I have chosen, but the same questions obviously need to be answered for other experimental agents too.
134. Cooper, D. A., Gatell, J. M., Kroon, S. Clumeck, N., Millard, J., Goebel, F. D., Bruun, J. N., Stingl, G., Melville, R. L., González-Lahoz, J. et al. Zidovudine in Persons with Asymptomatic HIV Infection and CD4+ Cell Counts Greater Than 400 per Cubic Millimeter. *New England Journal of Medicine*, 1993; 329: 297-303.
135. Anonymous. Cruel Confusion. *New Scientist*, August 7, 1993: 3.
136. See Brown, P. Wellcome under Fire over "Old" AZT Data. *New Scientist*, August 7, 1993: 4.
137. As quoted in Brown, P. Wellcome under Fire, 1993: 4.
138. Ellenberg, S. Discussion of "Surrogate Marker in AIDS and Cancer Trials." *Statistics in Medicine*, 1994; 13: 1439.
139. Schlumpberger, J. M., Wolde-Tsadik, G., Yao, J. F. and Hara, J. CD8+ Lymphocyte Counts and the Risk of Death in Advanced HIV Infection. *Journal of Family Practice*, 1994; 38(1): 33-8. (Emphasis added.)
140. Zaretsky, M. D. AZT Toxicity and AIDS Prophylaxis: Is AZT Beneficial for HIV+ Asymptomatic Persons with 500 or More T4 Cells Per Cubic Millimeter? *Genetica*, 1995; 95: 91-101.
141. Deer, B. AIDS Doctors Attack Drug Claims. *Sunday Times* [London], August 1, 1993: A1.
142. Phillips, A. N. and Saban, C. A. Zidovudine in Asymptomatic HIV Infection. *New England Journal of Medicine*, 1993; 329: 1895.
143. Berger, P. B. Zidovudine in Asymptomatic HIV Infection. *New England Journal of Medicine*, 1993; 329: 1895.
144. This problem has eased a bit in the United States since 1987, when the FDA began to allow patients with terminal diseases to use so-called investigational new drugs (IND) on the basis of physicians' prescription, under the following premises: "(1)The drug is intended to treat a serious or immediately life-threatening disease; (2) there is no comparable or satisfactory alternative drug or other therapy available to treat that stage of the disease in the intended patient population; (3) the drug is under investigation in a controlled clinical trial under an IND in effect for the trial, or all trials have been completed; and (4) the sponsor

of the controlled clinical trial is actively pursuing marketing approval of the investigational new drug with due diligence." Criteria as quoted in: Annas, G. J. Faith (Healing), Hope, and Charity, 1990: 188. Similar solutions are still nonexistent in most Western European countries. Hence, my reservations still refer to real situations and jurisdictions outside the United States. Apart from this, the American solution remains unsatisfactory, if my argument succeeds, because a necessary condition is that "the drug is under investigation in a controlled clinical trial under an IND in effect for the trial, or all trials have been completed." This still excludes the majority of drugs people with AIDS purchase illegally in the vast drug underground, which has turned into a multimillion dollar industry in the last decade.

145. Nussbaum, B. *Good Intentions*, 1990: 217.
146. Nussbaum, B. *Good Intentions*, 1990: 192-193.
147. Arras, J. D. Noncompliance in AIDS Research, 1990: 29.
148. Engelhardt, T. H.. Health Care Allocations: Responses to the Unjust, the Unfortunate, and the Undesirable. In Shelp, E. E. (Ed.). *Justice and Health Care*. Dordrecht, Netherlands: D. Reidel Publishing, 1981: 127.
149. I am not advocating that people should disobey the law. I merely wish to point out that injustice exists in a society where disobeying the law is easier for more affluent citizens than it is for the poor.
150. A recent example that I have come across was an advertisement of Melbourne's Fairfield Hospital published in a gay magazine in this city. The researchers offered interleukin-2 (one of the drugs many people with AIDS were, at that time, keen to try) to prospective participants. (In regard to the value of interleukin, see my comments earlier in this chapter). The exclusion criteria excluded not only people with more or less than 200-500 CD4+ cells, those who have not been on antiviral drugs, those who use alternative treatments such as bitter melon, compound Q, as well as therapeutic vaccines, but also people using illicit drugs such as heroin or speed. Finally, pregnant or lactating women were excluded from this particular trial. I do not wish to repeat here the problems mentioned in the main text, which are inherent in a strategy such as this, for the predictive value of this trial's outcome. It seems clear, however, that not only a large segment of the patient population is excluded from this trial, but also that it seems to be designed for a certain type of gay middle-class patient (believed to cause no troubles to researchers over the run of the trial). See: Victorian AIDS Council/Gay Men's Health Centre. Interleukin-2 Trial. *Melbourne Star Observer*, 1995; 251 (suppl. Positive Living): 7.
151. Of course, it could be argued that there is still a difference between the clinical researchers, on the one hand, who are not responsible for the status quo, and the legislators who are. One could argue, furthermore, that the patient thus has a moral obligation to the principal investigator after all (to keep the promises he or she made when joining the trial) because the researcher cannot be held responsible for the current unjust situation. I find this argument flawed. The main reason for this is that, even though the principal investigator is not primarily responsible for the current legal situation, he or she has clearly volunteered to work under

such legislation (and therefore make it work in the first place). The researcher has decided to work in a field where terminally ill people are coerced into participating in the very trials that he or she sets up. The researcher is nothing other than the representative of the legislator (just as a policeman is the executive of the legislator), and patients are entitled to treat him or her as such.

152. Macklin, R. and Friedland, G. AIDS Research: The Ethics of Clinical Trials. *Law, Medicine and Health Care*, 1986; 14: 276.

153. Wagner, W. Ethik der Arzneimittelprüfung am Menschen. In Wagner, W. (Ed.). *Arzneimittel und Verantwortung—Grundlagen und Methoden der Pharmaethik*. Berlin: Springer Verlag, 1993:161.

154. In the context of this statement, it might be helpful to recall my earlier qualifications that specify the conditions under which the administration of a placebo is actually equal to nontreatment. In regard to the idea that the placebo should not have a lesser likelihood of success or failure than the experimental agent at the beginning of the trial, see the argument by the Hellmans, which I will discuss under the heading "Placebos and Other Drugs" further on in this chapter.

155. This remains true even if, as is estimated today, two out of three drugs turn out to be useless in phase III clinical trials. The reason for this is that a one in three chance is still a better option than a zero chance with a placebo. As to the question of the harm that ineffective drugs themselves might cause, I refer to my chapter on paternalistic strategies that defend the current situation of nonaccess to experimental drugs (in particular, in the paragraphs dealing with Robert Young's argumentation), where I deal with this matter at great length.

156. Arras, J. D. Noncompliance in AIDS Research, 1990: 27.

157. Arras, J. D. Noncompliance in AIDS Research, 1990: 25.

158. Freedman, B. Equipoise and the Ethics of Clinical Research. *New England Journal of Medicine*, 1987; 317: 141-145.

159. Hellman, S. and Hellman, D. S. Of Mice but Not Men, 1991: 1856.

160. Hellman, S. and Hellman, D. S. Of Mice but Not Men, 1991: 1586.

161. Hogan, C. Test Treatment Strategies under Real Conditions, 1995: 8.

162. See, for example, Kadane, J. B. Progress Toward a More Ethical Method for Clinical Trials. *Journal of Medicine and Philosophy*, 1986; 11: 385-404; Merigan, T. C. You Can Teach an Old Dog New Tricks: How AIDS Trials Are Pioneering New Strategies. *New England Journal of Medicine*, 1990; 323(19): 1341-1343; Sonnabend, J. A. Do We Need New Ways to Evaluate Experimental AIDS Treatments? *AIDS Forum*, 1989; 1(1): 4-8.

163. Ellenberg, S. Discussion of "Surrogate Markers in AIDS and Cancer Trials," 1994: 1439.

164. That is not to say that such trials can ultimately be a replacement for the randomized clinical trial. Perhaps, as David Byar and colleagues pointed out, well-designed and controlled randomized "clinical trials remain the most reliable method for evaluating the efficacy of therapies." Byar, D. P., Simon, R. M., Friedewald, W. T., Schlesselman, J. J., DeMets, D. L., Ellenberg, J. H., Gail, M. H., and Ware, J. H. Randomized Clinical Trials. *New England Journal of Medicine*, 1976; 295(2): 74.

Chapter 3

1. United Nations General Assembly. *Universal Declaration of Human Rights*. Adopted and proclaimed on December 10, 1948. New York: United Nations, 1976.
2. United Nations General Assembly. International Covenant on Economic, Social and Cultural Rights. Resolution 220 (XXI). Adopted on December 16, 1966 (A/RES/2200 [XXI]). Reprinted in *American Journal of International Law,* 1967; 61: 861-870. (Emphasis added.)
3. World Health Organization. *Basic Documents*, Twenty-Sixth Edition. Geneva: WHO, 1976: 1.
4. Häyry, M. and Häyry, H. Health Care as a Right, Fairness and Medical Resources. *Bioethics,* 1990; 4(1): 1-21.
5. Goodin, R. E. *Reasons for Welfare—The Political Theory of the Welfare State*. Princeton, NJ: Princeton University Press, 1988: 6-7.
6. King, D. S. and Waldron, J. Citizenship, Social Citizenship and the Defence of Welfare Provision. *British Journal of Political Science,* 1988; 18: 415-443. (All quotes excerpted from: 442.)
7. I use the term moral right within this book as follows: Whenever I speak of a moral right of A to X, I mean that A has the disposition to have, and has, in reality, interests and that A has an ethically justifiable interest to X. The moral right in this interpretation is normally accompanied by corresponding moral obligations of others to either help A to satisfy his or her interests or, at least, to not prevent A actively from trying to satisfy his or her legitimate interests.
8. For a good discussion of the problems involved in such concepts see: Nelson, J. O. Against Human Rights. *Philosophy,* 1990; 65(253): 341-348; also Jeremy Waldron's excellent reader *Nonsense upon Stilts: Bentham, Burke and Marx on the Rights of Man.* London and New York: Methuen, 1987.
9. Feinberg, J. "Harmless Immoralities" and Offensive Nuisances. In Feinberg, J. *Rights, Justice, and the Bounds of Liberty—Essays in Social Philosophy.* Princeton, NJ: Princeton University Press, 1980: 69-109. The quotes are taken from pages 91 and 92, respectively.
10. Zippelius, R. *Das Wesen des Rechts*, Fourth Edition. München, Germany: C. H. Beck, 1978: 50f.
11. Lee, P. R. and Emmott, C. Health-Care System. In Reich, W. T. (Ed.). *Encyclopedia of Bioethics*, Volume 2, New York: The Free Press, 1972: 610-619; Jonsen, A. R. Right to Health-Care Services. Reich, W. T. (Ed.). *Encyclopedia of Bioethics*, Volume 2. New York: The Free Press, 1972: 623-630. Branson, R. Health-Care: Theories of Justice and Health Care. In Reich, W. T. (Ed.). *Encyclopedia of Bioethics*, Volume 2, New York: The Free Press, 1972: 630-637.
12. See, for examples of such positions: Harris, R. N. Private Consensual Adult Behavior: The Requirement of Harm to Others in the Enforcement of Morality. *UCLA Law Review,* 1967; 14: 581-603; Schüklenk, U. AIDS—Beyond Education. *Medical Journal of Australia,* 1992; 157: 723.
13. United Nations, General Assembly. International Covenant, 1967.

14. Kant, I. *Groundworks of the Metaphysics of Morals.* New York: Harper & Row, 1964.

15. Mill, J. S. *Utilitarianism, Liberty, Representative Government.* London: Dent, 1960.

16. Dworkin, G. *The Theory and Practice of Autonomy.* Cambridge, MA: Cambridge University Press, 1988.

17. Rousseau, J. *The Contract Social and Discourses.* London: Dent, 1973: 177-178.

18. American Medical Association. *AMA House of Delegates Proceedings.* Twenty-Third Clinical Convention, November 30—December 3, 1969: 183.

19. Buchanan, A. and Brock, D. W. Deciding for Others. *Milbank Quarterly,* 1986; 64(2): 53.

20. Christiansen, J. B. A Supply-Side View of American Medicine: Structure, Conduct, and Performance Since 1970. In Libecap, G. D. (Ed.). *Advances in The Study of Entrepreneurship, Innovation, and Economic Growth,* Volume 4. Greenwich, CT: JAI Press, 1990.

21. Pence, G. E. *Classical Cases in Medical Ethics: Accounts of Cases That Have Shaped Medical Ethics, with Philosophical, Legal and Historical Background,* Second Edition. New York: McGraw-Hill, 1994: 447.

22. Faaland, J, Parkinson, J. R. and Saniman, R. *Growth and Ethnic Inequality—Malaysia's New Economic Policy.* New York: St. Martin's Press, 1990: 127-128.

23. Publicly funded health care in Australia does not cover costs incurred by the use of dentist or optometrist services. These are the only exceptions to the otherwise existing publicly funded health care.

24. Young, R. *Personal Autonomy: Beyond Negative and Positive Liberty.* Keat, OH: Croon Helm, 1986: 47.

25. This assumes that the average patient's treatment relies entirely on approved drugs and the patient's participation in clinical trials.

26. It seems as if the standard arguments against such a strategy (such as the claim that there is no methodologically acceptable evidence for the assertion that these alternative remedies have any positive impact on a given patient) are no longer universally accepted as sufficient. This can be most easily shown by looking into recent developments in the insurance industry itself, which is generally more concerned with scaling back insurance payouts. A recent article in a U.S. AIDS magazine, by and for people with AIDS, reports that large insurance companies have set up pilot programs covering services requested by large numbers of their members. These included, most notably, "homeopathy, acupuncture and naturopathy, as well as preventive programs such as nutritional counseling and stress management." Bob Lederer, the article's author, notes correctly that these are the types of "alternative health care that the insurance industry as a whole has long been fighting tooth and nail." Lederer, B. Why Activism Matters, Some Insurance Companies Are Covering Alternative Health Care. *POZ magazine,* December 1994/January 1995.

27. Singer, P. *Practical Ethics*, Second Edition. Cambridge, MA: Cambridge University Press, 1993: 21-26.
28. Singer, P. *Practical Ethics*, 1993: 21.
29. Daniels, N. *Just Health Care*. Cambridge, MA: Cambridge University Press, 1985: 154-155.
30. Mohr, R. D. *Gays/Justice: A Study of Ethics, Society and Law*. New York: Columbia University Press, 1988.
31. Illingworth, P. *AIDS and the Good Society*, London: Routledge, 1990: 143ff.
32. Mohr, R. D. *Gays/Justice*, 1988: 230.
33. Mohr, R. D. *Gays/Justice*, 1988: 236.
34. Mohr, R. D. *Gays/Justice*, 1988: 230.
35. Nozick, R. *Anarchy, State and Utopia*. New York: Basic Books, 1974: 153.
36. Nozick, R. *Anarchy, State and Utopia*, 1974: 160.
37. Beauchamp, T. L. and Childress, J. F. *Principles of Biomedical Ethics*, Fourth Edition. New York: Oxford University Press, 1994: 337.
38. Illingworth, P. *AIDS and the Good Society*, 1990: 145.
39. This is, of course, not to say that I would not prefer such a welfare-based system. I think, however, that for the case under consideration, it is sufficient to show that financial support for people with AIDS who wish to buy and use experimental agents can be made even for societies without universal health care.
40. Illingworth, P. *AIDS and the Good Society*, 1990: 148.
41. Illingworth, P. *AIDS and the Good Society*, 1990: 137.
42. Illingworth, P. *AIDS and the Good Society*, 1990: 157.
43. This inevitably brings with it the negative consequences that I have mentioned in Chapter 2.
44. National Centre in HIV Epidemiology and Clinical Research. *Australian HIV Surveillance Report* 1994; 10(1): 7, 15.
45. As an aside, it should be mentioned that AIDS activists have refused to support the setup of such databases because they feared the FDA would use such information to discredit alternative remedies. For instance, "ACT UP New York led a successful coalition effort to stop the National Institutes of Health from funding an insurance industry organized computer database on so-called evaluations of alternative treatments. Activists charged the project would be a blacklist based on superficial reports by biased researchers to justify rejecting reimbursements. The database was shelved." Lederer, B. Why Activism Matters, 1994/1995: A most recent exchange of letters between members of ACT UP New York City and Princeton University's Arnold Levine, the chair of the U.S. National Institutes of Health's AIDS Research Evaluation Working Group, demonstrates that these fears are still dominant. George M. Carter wrote a scornful letter to Levine criticizing him for dismissing a whole list of alternative uninvestigated therapeutic approaches that the ACT UP chapter in Manhattan had sent to Levine. Carter argued that it "behooves the NIH to examine these various modalities with greater rigor than it has to date. To merely dismiss them is arrogant."

Carter, G. M., Levine, A. science.medicine.aids. article no. 12282. Posted May 22, 1995, 13:39:24.

46. Hogan, C. Test Treatment Strategies under Real Conditions. Treatment Issues, 1995; 9(4): 7.

47. Schüklenk, U. Against Manipulative Campaigns by "Community Based" AIDS Organizations. *Health Care Analysis,* 1994; 2(3): 253-261.

48. I have to concede that we have no studies providing evidence that such information would stop people with AIDS from using drugs that are proven to be harmful. Given the knowledge we have about the therapeutic misconception, we have some (weak) reason to assume that some people with AIDS will continue buying and using drugs that have been proven to have an overall negative impact on their health and quality of life.

49. National Library of Medicine. Funding Announcement. science.medicine.aids article no. 11585. Posted March 22, 1995.

50. National Library of Medicine. Funding Announcement, 1995.

51. Anonymous. New Grant Funds Study of Natural Healing. *Delicious!,* January 1995: 12-13.

52. Root-Bernstein, R. *Rethinking AIDS: The Tragic Cost of Premature Consensus.* New York: The Free Press, 1993: 362-363.

53. Moore, R. D., Stanton, D., Gopalan, R., and Chaisson, R. E. Racial Differences in the Use of Drug Therapy for HIV Disease in an Urban Community. *New England Journal of Medicine,* 1994; 330(11): 763.

54. Ayanian, J. Z. Race, Class, and the Quality of Medical Care. *JAMA,* 1994; 271(15): 1207-1208.

55. Young, R. *Personal Autonomy.* 1986: 49.

Conclusion

1. If they did actually work, the disease would no longer be considered terminal and would, therefore, fall into a different category altogether.

2. This is an option that is widely supported as something that should be available to all people with AIDS, by Australian AIDS organizations such as the Australian Federation of AIDS Organizations, an umbrella organization of AIDS Councils, and such groups.

3. The hijacker may well be assumed to have similarly noble motives as those generally attributed to clinical researchers.

Bibliography

Abrams, D. I. On the Matter of Survival. *BETA: Bulletin of Experimental Treatments for AIDS*, 1992; November: 12-13.
Abrams, D. I. Survival: The Ultimate Endpoint. *AIDS/HIV Treatment Directory*, 1993, 6(2):5-11.
Abrams, D. I. Personal Communication. March 7, 1995.
Altman, D. *AIDS and the New Puritanism.* London: Photo Press, 1986.
Altman, D. Community-Based AIDS Organizations and the Future. *National AIDS Bulletin* (Australia), 1990; 4(1): 16-18.
Altman, L. Experts Change Guidelines to Using Drugs for HIV. *The New York Times*, June 27, 1993: A1.
American Medical Association. *AMA House of Delegates Proceedings.* Twenty-Third Clinical Convention, November 30-December 3, 1969.
Annas, G. Faith (Healing), Hope, and Charity at the FDA: The Politics of AIDS Drug Trials. In Gostin, Lawrence O. (Ed.). *AIDS and the Healthcare System.* New Haven, CT: Yale University Press, 1990: 183-194.
Anonymous. British Widow Sues U.S. Agency Over AZT. *The Reuters European Business Report*, January 28, 1994.
Anonymous. Cruel Confusion. *New Scientist*, August 7, 1993: 3.
Anonymous (comment made on-line). Maury Povish show on ozone therapy. science.medicine.aids poster no. 14334; April 27, 1994, 16:34:02 GMT.
Anonymous. Medical Matters: Evaluating Alternative Treatments. People with AIDS Coalition New York City *Newsline*, 1994; March: 25-27.
Anonymous. New Grant Funds Study of Natural Healing. *Delicious*, January 1995: 12-13.
Anonymous. Protein Seems to Stimulate AIDS-Damaged Immune Systems. *Reuters Newsagency Dispatch*, March 2, 1995.
Anonymous. Zidovudine Again. *Lancet*, 1992; 339: 1105-1106.
Anonymous. Zu wenig Probanden für AZT-Versuch. *Die Neue Ärztliche*, January 4, 1988.
Appelbaum, P. S., Roth, L. H., Lidz, C. W., Benson, P., and Winslade, W. False Hopes and Best Data: Consent to Research and the Therapeutic Misconception. *Hastings Center Report*, 1987; 17(2): 20-24.
Archard, D. For Our Own Good. *Australasian Journal of Philosophy*, 1994; 72(3): 283-293.
Aristotle. *Nicomachean Ethics.* Oxford: Oxford University Press, 1966.
Arky, R. et al. (Eds.). *Physicians Desk Reference*, Forty-Eighth Edition. Montvalle: Medical Economics Data Production Company, 1994.

Armington, K. Evaluating New or "Alternative Therapies." *Treatment Issues,* 1993; 7(11/12): 28-29.

Arneson, R. Mill versus Paternalism. *Ethics,* 1980; 90(4): 470-489.

Arras, J. D. Noncompliance in AIDS Research. *Hastings Center Report,* 1990; 20(5): 24-32.

Ayanian, J. Z. Race, Class, and the Quality of Medical Care. *JAMA,* 1994; 271(15): 1207-1208.

Badewitz-Dodd, L. H. (Ed.). *MIMS 1994 Annual* (Australian Edition). Crows Nest, Australia: MediMedia, 1994.

Baenen, L. Off AZT. *Associated Press,* May 16, 1994.

Bayer, R. *Homosexuality and American Psychiatry.* Princeton: Princeton University Press 1987.

Bayer, R. *Private Acts, Social Consequences. AIDS and the Politics of Public Health.* New Brunswick, NJ: Rutgers University Press, 1991.

Beauchamp, T. L. Paternalism and Bio-Behavioral Control. *The Monist,* 1976; 60: 62-80.

Beauchamp, T. L. and Childress, J. *Principles of Biomedical Ethics,* Fourth Edition. New York: Oxford University Press, 1994.

Beauchamp, T. L. and Walters, L. (Eds.). *Contemporary Issues in Bioethics.* Belmont, CA: Wadsworth Publishing, 1994.

Benn, S. I. Freedom, Autonomy, and the Concept of a Person. *Proceedings of the Aristotelian Society,* 1976; 76: 109-130.

Berger, P. B. Zidovudine in Asymptomatic HIV Infection. *New England Journal of Medicine,* 1993; 329: 1895.

Berita AIDS Pusat Perubatan. Newsletter published by University of Malaya. 1993; 1(1).

Birnbacher, D. *Verantwortung für zukünftige Generationen.* Stuttgart, Germany: Reclam Verlag, 1988.

Blazey, P. AZT: A Cure for Dandruff? *With Complements,* 1994; 3(?): 1-3.

Branson, R. Health-Care: Theories of Justice and Health-Care. Reich, Warren T. (Ed.). *Encyclopedia of Bioethics,* Volume 2. New York: The Free Press, 1972: 630-637.

Broad, W. and Wade, N. *Betrayers of the Truth: Fraud and Deceit in the Halls of Science.* New York: Simon and Schuster, 1982.

Brock, D. W. Decision Making Competence and Risk. *Bioethics,* 1991; 5(2): 105-112.

Brock, D. W. The Ideal of Shared Decision Making Between Physicians and Patients. *Kennedy Institute of Ethics Journal,* 1991; 1: 28-47.

Brock, D. W. and Wartman, S. A. When Competent Patients Make Irrational Choices. *New England Journal of Medicine,* 1990; 322(22): 1595-1599.

Brook, I. Approval of Zidovudine (AZT) for Acquired Immunodeficiency Syndrome: A Challenge to the Medical and Pharmaceutical Communities. *JAMA,* 1987; 258(11): 1517.

Brown, P. Wellcome under Fire over "Old" AZT Data. *New Scientist,* August 7, 1993: 4.

Brown, S. Study Says Wellcome's AZT Ineffective. *Reuter's Financial Report*, January 30, 1994.

Brudner, J. Peptide T Fails in Phase II Test. *Treatment Issues*, 1994; 8(1): 5-9.

Buchanan, A. E. and Brock, D. W. Deciding for Others. *Milbank Quarterly*, 1986; 64(2): 17-94.

Buchanan, A. E. and Brock, D. W. *Deciding for Others: The Ethics of Surrogate Decision Making*. Cambridge, MA: Cambridge University Press, 1990.

Buchholz, W. M. Competent Patients and Irrational Choices. *New England Journal of Medicine*, 1990; 323(19): 1354-1355.

Byar, D. P., Simon, R. M. Friedewald, W. T., Schlesselman, J. J., DeMets, D. L., Ellenberg, J. H., Gail, M. H., and Ware, J. H. Randomized Clinical Trials. *New England Journal of Medicine*, 1976; 295(2): 74-80.

Cameron, M. E. *Living with AIDS—Experiencing Ethical Problems*. Newbury Park, CA: Sage, 1993.

Carter, G. M. Stinkhorn. science.medicine.aids article no. 11833, April 18, 1995.

Carter, G. M. Levine. science.medicine.aids article no. 12282. Posted May 22, 1995, 13:39:24.

Centers for Disease Control. Opportunistic Non-Hodgkin's Lymphomas Among Severely Immunocompromised HIV-Infected Patients Surviving for Prolonged Periods on Antiretroviral Therapy—United States. *Morbidity and Mortality Weekly Report*, 1991; 40(34): 591-600.

Centers for Disease Control. 1991 Revised Classification System for HIV Infection and Expanded Surveillance Case Definition for AIDS Among Adolescents and Adults. *Morbidity and Mortality Weekly Report*, 1992; 41(RR-17): 1-19.

Chapman, S. AIDS—Beyond Education (Letter to the Editor). *Medical Journal of Australia*, 1992; 157: 501.

Childress, J. F. Mandatory HIV Screening and Testing. In Beauchamp, T. L. and Walters, L. (Eds.). *Contemporary Issues in Bioethics*. Belmont, CA: Wadsworth Publishing, 1994: 557-571.

Choi, S., Lagakos, S. W., Schooley, R. T., and Volberding, P. A. CD4+ Lymphocytes Are an Incomplete Surrogate Marker for Clinical Progression in Persons with Asymptomatic HIV Infection Taking Zidovudine. *Annals of Internal Medicine*, 1993; 118(9): 674-680.

Christiansen, J. B. A Supply-Side View of American Medicine: Structure, Conduct, and Performance Since 1970. In Libecap, G. D. (Ed.). *Advances in the Study of Entrepreneurship, Innovation, and Economic Growth*, Volume 4, Greenwich, CT: JAI Press, 1990.

Concorde Coordinating Committee. Concorde: MRC/ANRS Randomised Double-Blind Controlled Trial of Immediate and Deferred Zidovudine in Symptom-Free HIV Infection. *Lancet*, 1994; 341: 871-880.

Cooper, D. A., Gatell, J. M., Kroon S., Clumeck, N., Millard, J., Goebel, F. D., Bruun, J. N., Stingl, G., Melville, R. L., González-Lahoz, J. et al. Zidovudine in Persons with Asymptomatic HIV Infection and CD4+ Cell Counts Greater

Than 400 per Cubic Millimeter. *New England Journal of Medicine*, 1993; 329: 297-303.
Cooper, E.. Controlled Clinical Trials of AIDS Drugs: The Best Hope. *JAMA*, 1989; 261: 2445.
Cote, T. R. and Biggar, R. J. Does Zidovudine Cause Non-Hodgkin's Lymphoma? *AIDS*, 1995; 9(4): 404-405.
Cox, S. Personal Communication. February 24, 1994.
Crimp, D. AIDS: Cultural Analysis/Cultural Activism. Crimp, Douglas. (Ed.). *AIDS: Cultural Analysis/Cultural Activism.* Cambridge, MA: MIT Press, 1988: 3-16.
Crimp, D. (Ed.). *AIDS: Cultural Analysis, Cultural Activism.* Cambridge, MA: MIT Press, 1988.
Daniels, N. *Just Health Care.* Cambridge, MA: Cambridge University Press, 1985.
D'Aquila, R. T., Johnson, V. A., Welles, S. L., Japour, A. J., Kuritzkes, D. R., DeGruttola, V., Reichelderfer, P. S., Coombs, R. W., Crumpacker, C. S., Kahn, J. O. et al. Zidovudine Resistance and HIV-1 Disease Progression During Antiretroviral Therapy. *Annals of Internal Medicine*, 1995; 122(6): 401-408.
Darby, A., Blake, E., and Dow, S. AMA Push for States to Reject Euthanasia. *The Age* (Melbourne), May 26, 1995: A6.
Deer, B. AIDS Doctors Attack Drug Claims. *Sunday Times* (London), August 1, 1993.
DeGruttola, V., Wolfsohn, M., Fischl, M., and Tsiatis, A. Modeling the Relationship Between Survival and CD4+-Lymphocytes in Patients with AIDS and ARC. *Journal of AIDS*, 1993; 6: 359-365.
Delaney, M. The Case for Patient Access to Experimental Therapy. *Journal of Infectious Diseases*, 1989; 159(3): 416-419.
Devlin, P. D.. *The Enforcement of Morals.* London: Oxford University Press, 1965.
Dixon, J. Catastrophic Rights: Vital Public Interests and Civil Liberties in Conflict. In Overall, C. and Zion, W. P. (Eds.). *Perspectives on AIDS: Ethical and Social Issues.* Toronto: Oxford University Press, 1991: 122-137.
Dobson, P. Personal Communication. February 25, 1994.
Duesberg, P. H. AIDS: Acquired by Drug Consumption and other Noncontagious Risk Factors. *Pharmacology and Therapeutics*, 1992; 55: 201-277.
Dworkin, G. Paternalism. Wasserstrom, R. A. (Ed.). *Morality and the Law.* Belmont, CA: Wadsworth Publishing, 1971: 107-126.
Dworkin, G. *The Theory and Practice of Autonomy.* Cambridge, MA: Cambridge University Press, 1988.
Dworkin, R. *Taking Rights Seriously.* London: Duckworth, 1978.
Edgar, H. and Rothman, D. J. New Rules for New Drugs: The Challenge of AIDS to the Regulatory Process. *Milbank Quarterly*, 1990; 68 (suppl. 1): 111-142.
Editorial. On Paternalism and Autonomy. *Journal of Medical Ethics*, 1983; 9: 4.
Editorial. On Stopping a Trial Before Its Time. *Lancet*, 1993; 342: 1311-1312.
Egger, M., Neaton, J. D., Phillips, A. N., and Smith, G. D. Concorde Trial of Immediate versus Deferred Zidovudine. *Lancet*, 1994; 343: 1355.

Ellenberg, S. Discussion of "Surrogate Markers in AIDS and Cancer Trials." *Statistics in Medicine,* 1994; 13: 1437-1440.

Engelhardt, T. H., Jr. Health Care Allocation: Responses to the Unjust, the Unfortunate, and the Undesirable. In Shelp, E. E. (Ed.). *Justice and Health Care.* Dordrecht, Netherlands: D. Reidel Publishing, 1981: 121-137.

Ervolino, B. What's a Mother to Do? *The Bergen Record,* May 9, 1994: B01.

Faaland, J., Parkinson J. R., and Saniman, R. *Growth and Ethnic Inequality—Malaysia's New Economic Policy.* New York: St. Martin's Press, 1990.

Faden, R. R. and Beauchamp, T. L. *A Theory and History of Informed Consent.* New York: Oxford University Press, 1986.

Feinberg, J. Legal Paternalism. *Canadian Journal of Philosophy,* 1971; 1(1): 105-124.

Feinberg, J. *Rights, Justice, and the Bounds of Liberty—Essays in Social Philosophy.* Princeton, NJ: Princeton University Press, 1980.

Feinberg, J. "Harmless Immoralities" and Offensive Nuisances. In Feinberg, J. *Rights, Justice, and the Bounds of Liberty—Essays in Social Philosophy.* Princeton, NJ: Princeton University Press, 1980: 69-109.

Feinberg, J. *The Moral Limits of the Criminal Law: Harm to Others.* New York: Oxford University Press, 1984.

Feinberg, J. *The Moral Limits of the Criminal Law: Harm to Self.* New York: Oxford University Press, 1986.

Feldman, J. The French Are Different—French and American Medicine in the Context of AIDS. *Western Medical Journal,* 1992; 157: 345-349.

Feldman, J. *An American Anthropologist in a Paris Clinic: Ethical Issues in the French Doctor-Patient Relationship.* Unpublished manuscript. University of Illinois, Urbana, Medical Scholar's Program / Department of Anthropology, 1992.

Fischl, M. A., Richman, D. D., Grieco, M. H., Gottlieb, M. S., Volberding, P. A., Laskin, O. L., Leedom, J. M., Groopman, J. E., Mildvan, D., Schooley, R. T. et al. The Efficacy of Azidothymidine (AZT) in the Treatment of Patients with AIDS and AIDS-Related Complex. *New England Journal of Medicine,* 1987; 317(4): 185-191.

Flannery, E. J. Should It Be Easier or Harder to Use Unapproved Drugs and Devices? *Hastings Center Report,* 1986: 16(1): 17-23.

Fleming, T. R. Surrogate Markers in AIDS and Cancer Trials. *Statistics in Medicine* 1994; 13: 1423-1435.

Foot, P. The Problem of Abortion and the Doctrine of the Double Effect. In Steinbock, B. (Ed.). *Killing and Letting Die.* Englewood Cliffs, NJ: Prentice-Hall, 1980: 156-165.

Frankfurt, H. G. Freedom of the Will and the Concept of a Person. *Journal of Philosophy,* 1971; 68(1): 5-20.

Freedman, B. Equipoise and the Ethics of Clinical Research. *New England Journal of Medicine,* 1987; 317: 141-145.

Garrett, L. Battle on AIDS Drug. *Newsday,* September 6, 1994: 4.

Gelbert, R. D., Lenderking, W. R., Cotton, D. J., Cole, B. F., Fischl, M. A., Goldhirsch, A., and Testa, M. A. Quality-of-life Evaluation in a Clinical Trial of Zidovudine Therapy in Patients with Mildly Symptomatic HIV Infection. *Annals of Internal Medicine,* 1992; 116(12 pt 1): 961-966.

Gonsalves, G. Viral Assay Reimbursements. science.medicine.aids article no. 11755. Posted April 7, 1995.

Goodin, R. E. *Reasons for Welfare—The Political Theory of the Welfare State.* Princeton, NJ: Princeton University Press, 1988.

Goodin, R. E. The Ethics of Smoking. *Ethics,* 1989; 99: 574-624.

Gray, J. N., Lyons, P. M., Jr., and Melton, G. B. *Ethical and Legal Issues in AIDS Research.* Baltimore, MD: The Johns Hopkins University Press, 1995.

Greenberg, J. An Alternative Treatment Activist Manifesto. *Treatment Issues,* 1993; 7(11/12): 1-2.

Gust, I. Control of Hepatitis B in Australia. *Medical Journal of Australia,* 1992; 156: 819-21.

Hamburg, M. A. and Fauci, A. S. HIV Infection and AIDS: Challenges to Biomedical Research. In Gostin, L. O. (Ed.). *AIDS and the Health Care System.* New Haven CT: Yale University Press, 1990.

Hamilton, J. D., Hartigan, P. M., Simberkoff, M. S., Day, P. L., Diamond, G. R., Dickinson, G. M., Drusano, G. L., Egorin, M. J., George, W. L., Gordin, F. M. et al. A Controlled Trial of Early versus Late Treatment with Zidovudine in Symptomatic Human Immunodeficiency Virus Infection. *New England Journal of Medicine,* 1992; 326: 437-443.

Hare, R. M. Could Kant Have Been a Utilitarian? *Utilitas,* 1993; 5(1): 1-16.

Harris, R. N. Private Consensual Adult Behavior: The Requirement of Harm to Others in the Enforcement of Morality. *UCLA Law Review,* 1967; 14: 581-603.

Hart, H. L. A. *Law, Liberty and Morality.* Oxford: Oxford University Press, 1963.

Havlir, D. and Richman, D. D. Antiretroviral Therapy. *Current Opinion in Infectious Diseases,* 1995; 8: 66-73.

Häyry, H. *The Limits of Medical Paternalism.* London: Routledge, 1991.

Häyry, M. and Häyry, H. Health Care as a Right, Fairness and Medical Resources. *Bioethics,* 1990; 4(1): 1-21.

Hellman, S. The Patient and the Public Good. *Nature Medicine,* 1995; 1(5): 400-402.

Hellman, S. and Hellman, D. S. Of Mice But Not Men: Problems of the Randomized Clinical Trial. *New England Journal of Medicine,* 1991; 324: 1585-1589.

Herek, G. M. Hate Crimes Against Lesbians and Gay Men—Issues for Research and Policy. *American Psychologist,* 1989; 44: 948-955.

Herek, G. M. and Brill, K. T. (Eds.). *Hate Crimes—Confronting Violence Against Lesbians and Gay Men.* Newbury Park, CA: Sage, 1992.

Hodel, D. The Dietary Supplementary Health and Education Act of 1993 (S.784/H.R. 1709). *Treatment Issues,* 1993; 7(11/12): 29-31.

Hodel, D. FDA Reform Floated in DC. *Treatment Issues* 1995; 9(6): 6-7.

Hogan, C. A Professional Patient's Survival Guide. *PWALive,* 1994; 5(3). 19-21.

Hogan, C. The Prudent Pariah: How to Be a Problem Patient. The CBC and diff. *PWALive*, 1994; 5(3): 22-29.
Hogan, C. Test Treatment Strategies under Real Conditions. *Treatment Issues*, 1995; 9(4): 7-9.
Hogan, C. and Hodges, J. CPCRA Master Antiretroviral Protocol Strategy (MAPS), Draft #4. Unpublished manuscript. Minneapolis: University of Minnesota, Division of Biostatistics, School of Public Health, September 23, 1993: 1-20.
Hooker, W. *Physician and Patient: or a Practical View of the Mutual Duties, Relations and Interests of the Medical Profession and the Community*. New York: Baker and Scribner, 1849.
Hutton, J. L. The Ethics of Randomised Controlled Trials: A Matter of Statistical Belief? *Health Care Analysis*, 1996; 4(2): 95-102.
Illingworth, P. *AIDS and the Good Society*. London: Routledge, 1990.
Illingworth, P. Warning: AIDS Health Promotion Programs May Be Hazardous to Your Autonomy. In Overall, C. and Zion, W. P. (Eds.). *Perspectives on AIDS: Ethical and Social Issues*. Toronto: Oxford University Press, 1991: 138-154.
Jonas, H. Philosophical Reflections on Experimenting with Human Subjects. *Daedalus*, 1969; 98: 219-247.
Jonsen, A. R. Right to Health-Care Services. In Reich, W. T. (Ed.). *Encyclopedia of Bioethics*, Volume 2. New York: The Free Press, 1972: 623-630.
Kadane, J. B. Progress Towards a More Ethical Method for Clinical Trials. *The Journal of Medicine and Philosophy*, 1986; 11: 385-404.
Kant, I. *Kritik der Praktischen Vernunft*. Hamburg, Germany: Felix Meiner Verlag, 1929.
Kant, I. *Lectures on Ethics*. London: Methuen, 1930.
Kant, I. *The Doctrine of Virtue*. New York: Harper and Row, 1964.
Kant, I. *Groundwork of the Metaphysics of Morals*. New York: Harper & Row, 1964.
Kant, I. *Critique of Practical Reason*. New York: Garland, 1976.
Kant, I. *The Metaphysics of Morals*. Cambridge, MA: Cambridge University Press, 1991.
King, D. S. and Waldron, J. Citizenship, Social Citizenship and the Defence of Welfare Provision. *British Journal of Political Science*, 1988; 18: 415-443.
Kirby, M. Informed Consent: What Does It Mean? *Journal of Medical Ethics*, 1983; 9(2): 69-75.
Kleinig, J. *Paternalism*. Totowa, NJ: Rowman and Allanheld, 1984.
Kobasa, S. C. Ouellette. AIDS and Volunteer Associations: Perspectives on Social and Individual Change. *Milbank Quarterly*, 1990; 68 (suppl. 2): 280-294.
Komrad, M. S. A Defence of Medical Paternalism: Maximising Patient's Autonomy. *Journal of Medical Ethics*, 1983; 9: 38-44.
Kovacs, J. A., Baseler, M., Dewar, R. J., Vogel, S., Davey, R. T., Falloon, J., Polis, M. A., Walker, R. E., Stevens, R., Salzman, N. P. et al. Increases in CD4+ Lymphocytes with Intermittent Courses of Interleukin-2 in Patients with HIV Infection. *New England Journal of Medicine*, 1995; 332: 567-575.

Kuhse, H. and Young, R. Options for Dying with Dignity. *Health Issues*, September 1987: 8-9.
Kwitny, J. *Acceptable Risks*. New York: Poseidon Press, 1992.
Lauritsen, J. *The AIDS War*, Second Edition. New York: Asklepios Press, 1993.
Lee, F. R. U. S. Cuts Three of Four AIDS Research Grants for Poor New Yorkers. *New York Times*, November 8, 1994: A1, B7.
Lee, P. R. and Emmott, C. Health Care System. In Reich, W. T. (Ed.). *Encyclopedia of Bioethics*, Volume 2. New York: The Free Press, 1972: 610-619.
Lederer, B. Why Activism Matters, Some Insurance Companies are Covering Alternative Health Care. *POZ Magazine*, December 1994/January 1995.
Lenderking, W. HIV-Infection and Long-Term Morbidity. *New England Journal of Medicine*, 1994; March 18.
Lenderking, W. R., Gelber, R. D., Cotton, D. J., Cole, B. F., Goldhirsch, A., Volberding, P. A. and Testa, M. A. Evaluation of the Quality of Life Associated with Zidovudine Treatment in Asymptomatic Human Immunodeficiency Virus Infection. *New England Journal of Medicine*, 1994; 330(11): 738-743.
Levine, C. and Bayer, R. The Ethics of Screening for Early Intervention in HIV Disease. *American Journal of Public Health*, 1989; 79: 1661-1667.
Levine, C., Neveloff Dubler, N., and Levine, R. J. Building a New Consensus: Ethical Principles and Policies for Clinical Research on HIV/AIDS. *IRB: A Review of Human Subjects Research*, 1991; 13(1-2): 1-17.
Libecap, G. D. (Ed.). *Advances in the Study of Entrepreneurship, Innovation, and Economic Growth*, Volume 4. Greenwich, CT: JAI Press, 1990.
Lim, L.C.C. Present Controversies in the Genetics of Male Homosexuality. *Annals Academy of Medicine Singapore*, 1995; 24: 759-762.
Macklin, R. and Friedland, G. AIDS Research: The Ethics of Clinical Trials. *Law, Medicine and Health Care*, 1986; 14: 273-280.
Marquis, D. An Argument that all Prerandomized Clinical Trials Are Unethical. *The Journal of Medicine and Philosophy*, 1986; 11: 367-383.
McLeod, G. X. and Hammer, S. M. Zidovudine: Five Years Later. *Annals of Internal Medicine*, 1992; 117: 487-501.
Melton, G. B., Levine, R. J., Koocher, G. P., Rosenthal, R., and Thompson, W. C. Community Consultation in Socially Sensitive Research: Lessons from Clinical Trials for Treatments of AIDS. *American Psychologist*, 1988; 43(7): 573-581.
Menzel, P. T. *Medical Costs, Moral Choices: A Philosophy of Health Care Economics in America*. New Haven and London: Yale University Press, 1983.
Merigan, T. C. You Can Teach an Old Dog New Tricks: How AIDS Trials Are Pioneering New Strategies. *New England Journal of Medicine*, 1990; 323(19): 1341-1343.
Merton, V. Community-Based AIDS Research. *Evaluation Review*, 1990; 14(5): 502-537.
Mill, J. S. *Utilitarianism, Liberty, Representative Government,*. London: Dent, 1960.

Mill, J. S. *Principles of Political Economy.* Harmondsworth, England: Penguin, 1970.
Mirken, B. AIDS Clinical Trials: Why They Have Recruiting Problems. *AIDS Treatment News,* 1995; 217.
Mohr, R. D. AIDS, Gays, and State Coercion. *Bioethics,* 1987; 1(1): 35-50.
Mohr, R. D. *Gays/Justice: A Study of Ethics, Society and Law.* New York: Columbia University Press, 1988.
Moore, R. D., Stanton, D., Gopalan, R., and Chaisson, R. E. Racial Differences in the Use of Drug Therapy for HIV Disease in an Urban Community. *New England Journal of Medicine,* 1994; 330(11): 763-768.
National Cancer Institute. RFA: CA-95-009. P.T. 34; K.W. 0715008, 0715035, 0755015. *NIH Guide,* 1995; 24(7).
National Centre in HIV Epidemiology and Clinical Research. *Australian HIV Surveillance Report,* 1994; 10(1).
National Gay and Lesbian Task Force Policy Institute. Anti-Gay/Lesbian Violence, Victimization, and Defamation in 1991. Washington, DC: NGLTF, 1992.
National Library of Medicine. Funding Announcement. science.medicine.aids article no. 11585. March 22, 1995.
Neaton, J. D., Wentworth, D. N., Rhame, F., Hogan, C., Abrams, D. I., and Deyton, L. Considerations in Choice of a Clinical Endpoint for AIDS Clinical Trials. *Statistics in Medicine,* 1994; 13(19-20): 2107-2125.
Nelson, J. O. Against Human Rights. *Philosophy,* 1990; 65(253): 341-348.
NGLTF Policy Institute. *Anti-Gay/Lesbian Violence, Victimization and Defamation in 1991.* Washington, DC: NGLTF, 1992.
Novick, A. Reflections on a Term of Public Service with the FDA Anti-Virals Advisory Committee. *Aids and Public Policy Journal,* 1993; 8(2): 55-61.
Nozick, R. *Anarchy, State, and Utopia.* New York: Basic Books, 1974.
Nussbaum, B. *Good Intentions: How Big Business and the Medical Establishment Are Corrupting the Fight Against AIDS, Alzheimer's, Cancer and More.* New York: Penguin, 1990.
Okie, S. Higher Cancer Rate Seen In Users of AIDS Drug. *The Washington Post,* August 16, 1990.
O'Neill, J. Some Good News About AIDS. *The Independent Monthly,* 1993; October: 28-32.
Osmond, D., Charlebois, E., Lang, W., Shoboski, S., and Moss, A. Changes in AIDS Survival Time in Two San Francisco Cohorts of Homosexual Men, 1983-1993. *JAMA,* 1994; 271(14): 1083-1087.
Overall, C. and Zion, W. P. (Eds.). *Perspectives on AIDS: Ethical and Social Issues.* Toronto: Oxford University Press, 1991.
Palca, J. Putting AIDS Research into Perspective. *Science,* 1991; 215: 1171.
Parfit, D. *Reasons and Persons.* Oxford: Clarendon Press, 1989.
Paton, H. J. Analysis of the Argument. In Kant, I. *Groundwork of the Metaphysics of Morals.* New York: Harper & Row, 1964.

Pence, G. E. *Classical Cases in Medical Ethics: Accounts of Cases That Have Shaped Medical Ethics, with Philosophical, Legal, and Historical Background*, Second Edition. New York: McGraw-Hill, 1995.

Phillips, A. N. and Sabin, C. A. Zidovudine in Asymptomatic HIV Infection. *New England Journal of Medicine*, 1993; 329: 1895.

Pluda, J. M., Yarchoan, R., Jaffe, E. S., Feuerstein I. M., Solomon, D., Steinberg, S., Wyvill, K. M., Raubitschek, A., Katz, D., and Broder, S. Development of Non-Hodgkins Lymphoma in a Cohort of Patients with Severe Human Immunodeficiency Virus (HIV) Infection on Long-Term Antiretroviral Therapy. *Annals of Internal Medicine*, 1990; 113: 276-282.

Rainbolt, G. W. Prescription Drug Laws: Justified Hard Paternalism. *Bioethics*, 1989; 3(1): 45-58.

Rainbolt, G. W. Justified Hard Paternalism: A Response to Ten. *Bioethics*, 1989; 3(2): 140-141.

Rawls, J. *A Theory of Justice*. Cambridge, MA: Harvard University Press, 1971.

Regan, D. H. Justifications for Paternalism. In Pennock, J. R. et al. (Eds.). *The Limits of Law*. New York: Lieber-Atherton, 1974: 189-210.

Richman, D. D., Fischl, M. A., Grieco, M. H., Gottlieb, M. S., Volberding, P. A., Laskin, O. L., Leedom, J. M., Groopman, J. E., Mildvan, D., Hirsh, M. S. et al. The Toxicity of Azidothymidine (AZT) in the Treatment of Patients with AIDS and AIDS-Related Complex. *New England Journal of Medicine*, 1987; 314(4): 192-197.

Rodwin, M. A. *Medicine, Money, and Morals*. New York: Oxford University Press, 1993.

Root-Bernstein, R. *Rethinking AIDS: The Tragic Cost of Premature Consensus*. New York: The Free Press, 1993.

Rothman, K. J. and Michels, K. B. The Continuing Unethical Use of Placebo Controls. *New England Journal of Medicine*, 1994; 331(6): 394-398.

Rousseau, J. *The Contract Social and Discourses*. London: Dent, 1973.

Salisbury, D. A. and Schechter, M. T. AIDS Trials, Civil Liberties and the Social Control of Therapy: Should We Embrace New Drugs with Open Arms? *Canadian Medical Association Journal*, 1990; 142(10): 1057-1062.

Salzman, B. Editorial. People with AIDS Coalition *Newsline* (New York City), 1993; 84: 3,15.

Sande, M. A., Carpenter, C. C. J., Cobbs, G., Holmes, K. K., and Sanford, J. P. (for the U.S. National Institute of Allergy and Infectious Diseases State-of-the-Art Panel on Anti-Retroviral Therapy for Adult HIV-Infected Patients). Antiretroviral Therapy for Adult HIV-Infected Patients: Recommendations from a State of the Art Conference. *JAMA*, 1993; 270(21): 2583-2589.

Schafer, A. Experimentation with Human Subjects: A Critique of the Views of Hans Jonas. *Journal of Medical Ethics*, 1983; 9: 76-79.

Schlumpberger, J. M., Wolde-Tsadik, G., Yao, J. F., and Hara, J. CD8+ Lymphocyte Counts and the Risk of Death in Advanced HIV Infection. *Journal of Family Practice*, 1994; 38(1): 33-38.

Schopenhauer, A. *On the Basis of Morality.* Indianapolis, IN: Bobbs-Merrill, 1965.
Schopenhauer, A. *Preisschrift über das Fundament der Moral.* Hamburg, Germany: Felix Meiner Verlag, 1979.
Schüklenk, U. AIDS—Beyond Education. *Medical Journal of Australia,* 1992; 157: 723.
Schüklenk, U. HIV-Related Phraseology. *Australian Journal of Public Health,* 1992; 16(3): 329-330.
Schüklenk, U. AIDS and the Bioethics Debate: Reading Abstracts Is Not Enough. In Joseph, K. (Ed.). *Conference Proceedings: Philosophy and Applied Ethics Re-Examined.* Newcastle, Australia: Australian Association for Professional and Applied Ethics, 1993: 145-160.
Schüklenk, U. Naturwissenschaften und Philosophie. In Lautmann, R. (Ed.). *Homosexualität. Handbuch der Theorie- und Forschungsgeschichte.* Frankfurt, Germany: Campus Verlag, 1993: 307-317.
Schüklenk, U. Against Manipulative Campaigns by "Community Based" AIDS Organizations. *Health Care Analysis,* 1994; 2(3): 253-261.
Schüklenk, U. AIDS and the Lab Rats. *Science and Public Affairs,* 1996; winter: 54-57.
Schüklenk, U. Ethische Probleme des Designs und der Zugangsvoraussetzungen klinischer AIDS Versuchsreihen. *Ethik in der Medizin,* 1997; 9(1): 15-30.
Schüklenk, U. and Hogan, C. Patient Access to Experimental Drugs and AIDS Clinical Trial Designs: Ethical Issues. *Cambridge Quarterly of Health Care Ethics,* 1996; 5: 400-409.
Schüklenk, U and Mertz, D. Christliche Kirchen und AIDS. In Dahl, E. (Ed.). *Die Lehre des Unheils—Fundamentalkritik am Christentum.* Hamburg, Germany: Carlsen Verlag, 1993: 263-279, 309-312.
Schüklenk, U, Mertz, D., and Richters, J. The Bioethics Tabloids: How Professional Ethicists Have Fallen for the Myth of Tertiary Transmitted Heterosexual AIDS. *Health Care Analysis,* 1995; 3(1): 27-36.
Schüklenk, U. and Riley, T. Homosexuality, Societal Attitudes Toward. In Chadwick, R. (Ed.). *Encyclopedia of Applied Ethics,* Volume 2. San Diego, CA: Academic Press, 1998: 597-603.
Schüklenk, U and Ristow, M. The Ethics of Research into the Causes of Homosexuality. *Journal of Homosexuality,* 1996; 31(3): 5-30.
Shorr, A. F. AIDS and the FDA: The Ethical Case for Limiting Patient Access to New Medical Therapies. *IRB: A Review of Human Subjects Research,* 1992; 14(4): 1-5.
Singer, P. *Practical Ethics,* Second Edition. Cambridge, MA: Cambridge University Press, 1993.
Skene, L. Risk-Related Standard Inevitable in Assessing Competence. *Bioethics,* 1991; 5(2): 113-117.
Smart, J. J. C. and Williams, B. *Utilitarianism: For and Against.* Cambridge, MA: Cambridge University Press, 1973.

Smart, J. J. C. An Outline of a System of Utilitarian Ethics. In Smart, J. J. C. and Williams, B. *Utilitarianism: For and Against.* Cambridge, MA: Cambridge University Press, 1973: 3-74.

Sonnabend, J. A. Do We Need New Ways to Evaluate Experimental AIDS Treatments? *AIDS FORUM,* 1989; 1(1): 4-8.

Sonnabend, J. A. Review of AZT Multicenter Trial—Data Obtained Under the Freedom of Information Act by Project Inform and ACT UP. *AIDS FORUM,* 1989; 1(1): 9-15.

Spiers, H.. Community Consultation and AIDS Clinical Trials (III). *IRB: A Review of Human Subjects Research,* 1991; 13(5): 3-7.

Sullivan, L. AIDS—Beyond Education. *Medical Journal of Australia,* 1992; 157: 286.

Tännsjö, T. The Morality of Clinical Research: A Case Study. *Journal of Medicine and Philosophy,* 1994; 19(1): 7-21.

Ten, C. L. Paternalism and Levels of Knowledge: A Comment on Rainbolt. *Bioethics,* 1989; 3(2): 135-139.

Ten, C. L. *Mill on Liberty.* Oxford: Oxford University Press, 1980.

Thompson, D. The AIDS Political Machine. *Time,* January 22, 1990: 24-25.

TRT-5. Concerning a Clinical Phase III trial of Saquinavir: Balancing Research and Health Care. science.medicine.aids article no. 11418. Forwarded by Mark Buzza. Posted February 22, 1995.

United Nations General Assembly. International Covenant on Economic, Social and Cultural Rights. Resolution 2200 (XXI). Adopted on December 16, 1966 (A/RES/2200 [XXI]). Reprinted in *American Journal of International Law,* 1967; 61: 861-870.

United Nations General Assembly. *Universal Declaration of Human Rights.* Adopted and proclaimed on December 10, 1948. New York: United Nations, 1976.

Veatch, R. M. Should I Enroll in a Randomized Clinical Trial? *IRB: A Review of Human Subjects Research,* 1988; 10(5): 7-8.

Victorian AIDS Council/Gay Men's Health Centre. Not All Clinical Trials Go Horribly Wrong. *Melbourne Star Observer,* 1995; 251 (suppl. Positive Living): 8.

Victorian AIDS Council/Gay Men's Health Centre. Interleukin-2 Trial. *Melbourne Star Observer,* 1995; 251 (suppl. Positive Living): 7.

Volberding, P. A. Is It Possible to Prove a Survival Benefit from Early Treatment? *Beta: Bulletin of Experimental Treatment for AIDS,* 1992; November: 12-15.

Wagner, W. Ethik der Arzneimittelprüfung am Menschen. In Wagner, W. (Ed.). *Arzneimittel und Verantwortung—Grundlagen und Methoden der Pharmaethik.* Berlin: Springer Verlag, 1993: 151-186.

Wagner, W. (Ed.). *Arzneimittel und Verantwortung—Grundlagen und Methoden der Pharmaethik.* Berlin: Springer Verlag, 1993.

Waldron, J. (Ed.). *Nonsense upon Stilts: Bentham, Burke and Marx on the Rights of Man.* London and New York: Methuen, 1987.

Wicclair, M. R. Patient Decision-Making Capacity and Risk. *Bioethics,* 1991; 5(2): 91-104.
Wilson, T. and Wilson, D. (Eds.). *The State and Social Welfare: The Objectives of Policy.* London and New York: Longman, 1991.
Wolfstädter, H. D. (Ed.). *Unkonventionelle Medizin bei HIV und AIDS.* Berlin: DAH (Deutsche AIDS Hilfe), 1995.
Wollheim, R. Crime, Sin and Mr. Justice Devlin. *Encounter,* 1959; 13: 34-40.
Wollheim, R. John Stuart Mill and the Limits of State Coercion. *Social Research,* 1973; 40(1): 1-30.
World Health Organization. *Basic Documents,* Twenty-Sixth Edition. Geneva: WHO, 1976.
Yarchoan, R., Hiroaki, M., Myers, C. E., and Broder, S. Clinical Pharmacology of 3',-Azido-2',3'-Dideoxythymidine (Zidovudine) and Related Dideoxynucleosides. *New England Journal of Medicine,* 1989; 321: 726-738.
Young, R. *Personal Autonomy: Beyond Negative and Positive Liberty.* Kent, OH: Croom Helm, 1986.
Zaretsky, M. D. AZT Toxicity and AIDS Prophylaxis: Is AZT Beneficial for HIV+ Asymptomatic Persons with 500 or More T4 Cells per Cubic Millimeter? *Genetica,* 1995; 95: 91-101.
Zippelius, R. *Das Wesen des Rechts,* Fourth Edition. München, Germany: C. H. Beck, 1978.

Index

Page numbers followed by the letter "t" indicate tables.

A posteriori knowledge, 16
A priori knowledge, 16
Abrams, Donald I.
  on Alpha trial, 117
  on AZT survival rate, 115,116
  on surrogate markers, 127
Acceptability, intervention and, 73
Access
  limited, 134
  restriction on drugs, 89-97
ACT UP, AIDS awareness, 2
Act-autonomy, 77
ACTGO19 trial, 123-124
Adverse reaction reports,
  Fischl/Richman trial,
  111,112-113,112t
Aerosolized pentamidine,
  69,108-109
Agence Nationale de Recherche
  sur le Sida, 109
AIDS
  communities, support for, 7
  databases, need for, 163
  experimental drugs and, 1
  minimum options, 136
  self-inflicted disease, 90
  syndrome, 93
AIDS Related Complex (ARC),
  107,116
AIDS-related dementia, 27
AIDS Treatment News, 96
Alpha trial, 117
Alternative medicine, 44-45,164-165
Alternative treatment activist, 67

Altruism, 32-33,98-100
American Civil Liberties Union,
  homosexual rights, 9
American Medical Association,
  on medical care rights, 150
Ancient Greece, autonomy in, 10
Annals of Internal Medicine, AZT
  resistance, 124
Annas, George J.
  on drug access, 91
  on experimental drug use, 49
Anti-Viral Drug Advisory
  Committee, FDA, 4
Appelbaum, Paul S., 97
Archard, David, on nondisclosure, 60
Aristotle, on self-harm, 24
Armington, Kevin, on alternative
  therapies, 67-68
Arneson, Richard, 38,46
Arras, John
  on access, 135,137
  on altruistic behavior, 99
  on promises, 138
  on society's rights, 93
Australia
  AIDS treatment, 106
  legalized euthanasia, 82
  PWA's in, 162
  suicide legality in, 9-10
  universal health care, 150,151
Authenticity, 43-44
  choice and, 84-85
  definition of, 57
  treatment choice, 45-46,85

219

Authenticity *(continued)*
 views on, 75,85
Autonomy
 Beauchamp and Childress on, 30
 Beauchamp, Childress, and Faden on, 64-74
 character ideal, 13,74
 concepts of, 42
 dispositional, 46,76-81
 Dworkin's definition of, 56-57
 elements of, 65-67
 hierarchy of issues, 58,72
 individual characteristics of, 26
 instrumental value of, 14,74-75
 issue of, 7-8
 marginalized populations and, 12,87
 maturity and, 27,29-30
 Mill, J. S., on, 11-12,23,24
 nondisclosure and, 61
 practical hindrances and, 167-168
 restrictions on, 77
 right of, 11
 suffering and, 27
 term, 42
 utility and, 30
 value of, 81
 virtues of, 43
 Western society, 10,11,14
Autos (self), 10
AZT
 approval of, 119,121-124
 discontinuance of use, 130
 unanswered questions about, 130-131
AZT clinical trials, 131
 ACTGO19, 123-124
 activist view of, 52-53
 Alpha, 117
 Fischl/Richman, 106-121
 French, 125-126
 legal actions, 119-121
 noncompliance in, 91
 problems with, 94-95

Bactrim, controlled trials on, 109
Bastyr University, alternative medicine catalog, 164
Beats (gay), 32
Beauchamp, Tom L.
 on autonomy, 30,64-74
 on Dworkin, 57
 on intervention, 32-33
 on justice, 159
 on omissions, 63
Belief systems, coherent, 84
Beneficence, 72
Brock, Dan
 on self-determination, 52
 standard of competence, 28-29,51-52,54-55
Brook, Itzhak, AZT trial termination, 121
Brown, Phyllida, 131
Buchanan, A. E., standard of competence, 28-29, 51-52,54-55
Burroughs-Wellcome pharmaceutical company, 113,121,131-132
Bush, George, deregulation policies, 7

Callen, Michael, 108
Canada, universal health care, 150
Cancer, experimental drugs and, 1
Carter, George M., on Stinkhorn, 69
Categorical Imperative, Kant's, 17,19,20
CD4+ lymphocytes
 information about, 59
 as surrogate marker, 116-117, 119,123-124,127,132,142
Chapman, Simon, 32
Cheating, clinical trials, 92,105,127,141
Children, incompetency in, 26
Childress, James F.
 on autonomy, 30,64-74

Childress, James F. *(continued)*
  on justice, 159
  on omissions, 63
Choice, moral duty and, 16
Christian moral philosophy, Kant and, 21
Clinical endpoints, criteria, 118-119
"Clinical equipoise," 139
Clinical research system, 135,140-141
Clinical trials
  ACTGO19 trial, 123-124
  Alpha, 117
  AZT, 94-95,131
  choice and, 89
  Concorde, 52,116,122, 124-125,131-132
  drop out problem, 92
  enrollment, 2,6,102,134-136
  experimental drugs and, 1
  Fischl/Richman, 106-121
  limited placement, 126
  morality of, 89
  noncompliance in, 91
  nucleoside analogue, 94
  patient withdrawal, 126-127
  physician's role in, 101-102
  placebo use in, 3,120,126,139-140
  purpose of, 97
  rationale for, 89
  risks of, 6,71
  trial design, 103,109-110
  voluntary participation in, 7,9,103
  wasting syndrome, 96,105-106
Compensatory justice, 158-159
Competence
  individual judgement on, 27,54
  liberty and, 9
  proof of, 29
  standard of, 28-29,51-52,54-55
Concorde trial, 52,116,122, 124-125,131-132
Consent, self-harm and, 24,31
Consequences, value judgements and, 18-19

Contradictio in adjecto, 21
Cooper, David, Concorde trial, 132
Cooper, Ellen
  on drug access, 90,96-97
  on patient enrollment, 103
Corti, Jim, 70
Cox, Spencer, on AIDS information, 70
Critique of Practical Reason,
  on moral duty, 16,18

Daniels, Norman, on lifestyle risks, 156
Dapsone, 110
DeGruttola, V., on CD4+ levels, 127
Delaney, Martin, 42
  on drug access, 93
  on noncompliance, 91-92
Dependency
  experimental drug use and, 6
  intervention and, 38
Deregulation, FDA, 4,7,123
Deutsche AIDS-Hilfe, 68
"Die-ins," 2
Diethylstilbestrol scandal, 4
Dispositional autonomy, 46,76-81,145
Dixon, John
  cheating in clinical trials, 92
  on drug access, 90
  on treatment control, 79
  trial design, 103
Drug approval, FDA, 3-4
  process, FDA, 5
Drug legalization, failure of, 13
Drug notification, FDA, 3
Drug scandals, regulatory power and, 4
Drug use, illegality of, 13
Dubs, Greg, 106
Duesberg, P. H., on lymphoma risk, 113-114
Dworkin, Gerald
  on experimental drug use, 64

Dworkin, Gerald *(continued)*
  on slavery, 38-39,57
  weak paternalism, 56-64

Elixir sulfanilamide scandal, 4
Ellenberg, Susan
  Concorde trial, 132
  on surrogate markets, 142
Emergency, paternalistic
  intervention, 58-59
Empirical interpretation, reason
  and, 83-84
Engelhardt, H. Tristam, 135
Engelman, Bradley, on euthanasia,
  82
Enrollment, clinical trials,
  102,134-136
Entitlement theory, 158
European Union, universal health
  care, 150
Euthanasia
  legalization of, 82
  support of, 80,81
Experimental agents, knowledge
  of, 66,71
Experimental drug use
  Annas' view on, 49
  costs of, 145-146,152-154
  dispositional autonomy and, 79,80
  Dworkin's view on, 64
  individual coverage, 156-161
  Mill's view of, 40
  requirements for, 87-88
  restricted access, 89-97
  right to choose, 82-85
  strong paternalism, 86
  supervised, 142-143
  universal health care, 151-156
  weak paternalism, 86
  Young's view of, 86,152
Experimental drugs
  efficacy of, 139-140
  information about, 161-165
  subsidizing, 155
  terminally ill and, 1

Faden, Ruth R.
  autonomy theory, 64-74
  on Dworkin, 57
Family, legal right to, 158-160
Fansidar, 110
Fauci, Anthony
  on AZT survival rate, 115
  legalization of aerosole
    pentamidine, 108-109
  on patient interest's, 95
Feasibility, intervention and, 73
Feinberg, Joel, weak paternalism,
  41-56
Feldman, Jamie, on deliberate
  omissions, 60-61,63
Fischl, Margaret, 109,111
Fischl/Richman AZT trial
  ethical analysis of, 121-133
  ethical implications of, 129
  patient interests in, 106-107,121
Flannery, Ellen J., 3
Fleming, Thomas R., on surrogate
  markers, 127
Food and Drug Administration
  (FDA)
  deregulation of, 4,7,123
  role of, 3-4
Foot, Philippa, 99-100
France, Bactrim research, 109
Frankfurt, Harry G., 56,75
Freedman, Benjamin, 139
Freedom, authenticity and, 76
Freedom of Information Act,
  Fischl/Richman trial, 110

"Gay bashing," 13
Gay bathhouses
  autonomy rights and, 11
  closing of, 32
Gay liberation movement, 13
Gay media, 45
Gay men, AIDS communities, 7
Gay-related immunodeficiency
  (GRID), 2
Genetica, on Concorde trial, 132-133

Germany
  AIDS information in, 165-166
  PWA's in, 162-163
  universal health care, 151
Gonsalves, Gregg, on viral load,
  117-118
Goodin, Robert
  on consent, 40
  on welfare state, 147
Greenberg, Jon, on alternative
  therapies, 67
*Groundwork of the Metaphysics
  of Morals*, 15,17

Hare, Richard M., on Kant, 20,23
Harm To Self, 50
"Harmless immorality," 148
Harris, Robert, on physical
  paternalism, 32
Hart, H. L. A, critique of, 32
Häyry, Heta
  decision-making knowledge, 66
  health care rights, 146-147
Häyry, Mati, health care rights,
  146-147
Health, moral right to, 147
Health care
  autonomy and, 57
  fundamental right of, 146
  individual coverage, experimental
    drug use in, 156-161
  marginalized populations and, 157
  universal coverage, experimental
    drug use in, 151-156
Health insurance, uninsured
  Americans, 150
Hellman, Deborah, on experimental
  drugs, 139
Hellman, Samuel
  on experimental drugs, 139
  on patient's rights, 95
Helmets, autonomy rights
  and, 11,35-36

Helsinki Declaration, patient's rights,
  105
Henninger, David, on FDA
  deregulation, 7
Heteronomous principles, 17,76
Hodel, Derek
  on AIDS health options, 70
  on FDA deregulation, 123
Hogan, Carlton, 68
  on NARTIs, 125
  on treatment options, 163-164
Homosexuality, illegality of, 13
Homosexuals
  AIDS and, 158-160
  AIDS communities, 7
  autonomy rights of, 11,12,35-36
  as marginalized population, 12-13
  right to self-determination, 9
Hooker, Worthington,
  on nondisclosure, 62
Humanistische Union, homosexual
  rights, 9

Illingworth, Patricia, 157-158,
  159-160
Immune status, disclosure of, 59
Inauthenticity, 43
Incompetence
  J. S. Mill on, 26
  paternalistic intervention, 58,59
Individual liberty, right of, 9
Individual rights, 95
Information, experimental drugs,
  146,151-156,165-167
Information overload, omissions
  and, 63-64
Informed consent, 27,40,95,126
  recall of, 55
Institution Review Board, FDA, 111
Intentionality, autonomy clement,
  65-66
Interleukin-2, 117
International Covenant on Economic,
  Social and Cultural Rights
  (1966), 146,149

Internet, science. medicine. aids, 52,69
Intervention
  altruistic, 32-33
  perception of, 42
  state's legitimate role, 42,51
  weak paternalism, 47
Interventions
  Dworkin's permissible, 58-60
  J. S. Mill on, 25,33-34
  state justified, 72-74
Intravenous drug users (IDUs)
  AIDS and, 160
  AIDS communities, 7
  marginalized population, 12,87
Ivory Coast, Bactrim research, 109

Jehovah's Witness, blood transfusion and, 74
Jonas, Hans, 98
Journal of the American Medical Association, AZT trial termination, 121
Journal of Infectious Diseases, 91
Judgement, Torbjörn Tännsjö on, 53
Justice, 72
"Justifications for Paternalism," 35

Kant, Immanuel, ethical theory of, 15-23
Killen, Jack Y., 106
King, Desmond S., 147
Kirby, Michael, consent forms, 55
Kleinig, John, passive paternalism, 15
Knowledge, types of, 16

Lancet, The
  on AZT approval process, 122,123-124
  on Fischl/Richman trial, 110
Large simple trials (LSTs), 53

Lauritsen, John, on Fischl/Richman trial, 110,112
Law, Kant's philosophy of, 19
Legal moralism, rejection of, 78
"Legal Paternalism," 46
Liberalism, definition of, 41
Libertarian theories, on justice, 159
Liberty
  J. S. Mill on limits of, 19-20
  Kant on limits of, 19
Life plan, autonomy and, 77,83

Maturity, relationship to autonomy, 27,29-30
Media, impact on authenticity, 45
Medical Journal of Australia, AIDS transmission prevention, 32
Medical paternalism, 38
Medical profession, homosexual attitude toward, 13
Mentally ill, incompetence in, 26
Metaknowledge, 66
Metaphysics of Morals, 19
Mill, John Stuart
  on autonomy, 11-12,23
  on Kant, 20
  philosophy of, 11-12,19-20,23,24-41
Mill on Liberty, 34
Mirken, Bruce, 96,105-106
Mohr, Richard D., 157-159
Moral agency, autonomy and, 75
Moral agent, Kant on, 16-17,20
Moral duty, Kantian, 15,16,18
Moral law, absolute, 16
Moral Limits of the Criminal Law, 41
Moral personality, Kant on, 19
Moral personhood, autonomy and, 75
Moral rights, rhetoric of, 148
Morality, as reflex, 46
Mortality, Fischl/Richman trial, 111,112,112t,115
Multiple-arm trials, 139

National Broadcasting Company
    (NBC), on Fischl/Richman
    trial, 110-111
National Institutes of Health (NIH),
    164
National Library of Medicine, 164
Natural lottery, 135
New England Journal of Medicine
    AIDS prophylactics use, 168
    on Concorde trial, 133
    Fischl/Richman trial,
        111,112,112t
    on zidovudine, 113
New Scientist, AZT prescription
    rules, 131
Newsline, AIDS treatment, 69
Nicomachean Ethics, 24
NIH Guide, on NHL, 114
Nomos (rule), 10
Noncompliance, clinical trials, 91-92
Noncontrolling influence, autonomy
    element, 65
Nondisclosure, arguments for, 60-63
Non-Hodgkin's lymphoma (NHL),
    AZT and, 113-114
Nonmaleficence, 72
Novick, Alvin
    on Anti-Viral Drug Advisory
        Committee, 4-5
    on clinical trials, 97-98
    on drop outs, 92
    substandard care, 104-105
Nozick, Robert, 158
Nucleoside analogue reverse
        transcriptase inhibitors
        (NARTIs), 125
    clinical trial, 94,128
Nussbaum, Bruce
    AIDS treatment information,
        69,109
    on AZT approval process, 122
Nutritional supplements, regulation
    of, 70

Offense principle, 148
Office of Alternative Medicine,
    NIH, 164
Officia iuris, 15
Officia virtutis, 15
Omissions
    deliberate, 60-61. *See also*
        Nondisclosure
    injury by, 36
On the Basis of Morality, 24
On Liberty, 11,24,38,41
Opportunistic infections, 94,95,
    108,111,127-128

"Passive paternalism," 15
Paternalism. *See also* Strong
        paternalism; Weak
        paternalism
    definition of, 58
    experimental drugs and, 1,6,8
Patient withdrawal, 126-127
Patient's rights, international, 105
Paton, H. J., 20
Peer pressure, AIDS treatment
    and, 71-72
Pence, Gregory E.
    on AZT therapy, 125
    United States health care, 150
Peptide T, use of, 47-48
Person, Kant's definition of, 19
Personal Autonomy, 75,152
Personality change, argument
    for paternalism, 35-36
Petitio principii, 17
Pharmaceutical advertising, 45
Pharmaceutical trusts, development
    of, 5
"Philosophical Reflections
        on Experimenting with
        Human Subjects," 98
Physician, clinical trial role, 101-102

Physicians Desk Reference,
  on Retrovir, 108,113,119
Placebos
  AZT trials, 91,126
  Fischl/Richman trial,
    107,110,111,126
  terminally ill and, 2-3,
    120,139-140
Pneumocystis carinii, 59,69,
    115,133,168
Poverty, experimental drug use
    and, 76,87-88
Practical Imperative, Kant's, 17-18
"Preventive treatment," 92
Principal investigators
  clinical trial role, 102-104
  enrollment role, 134-135
  view of clinical trials, 97
Principles of Biomedical Ethics, 63
Project Inform, 42
Promises, PWAs, 138
"Prudent Pariah, The," 68
PWALive magazine, 68

"Quacks," 3

Rainbolt, George, on drug
    prescriptions, 41
Rationality, standards of, 83-84
Rational-will utilitarian, 20
Reagan, Ronald
  deregulation policies, 7
  war against drugs, 13
Regan, Donald, argument
    for paternalism, 35-37,38
Religion
  rationality and, 83-84
  self-determination and, 14
Republic of Singapore
  homosexuals in, 13
  smoking in, 148-149
Research funds, GRID and, 2
Research interests, primacy of, 6-7

Responsibility, consequences and, 56
Retrovir (AZT), 108
Ribavirin, 69-70
Risk
  Aristotle on, 24
  Beauchamp and Childress on, 51
  competence and, 51-52
  as fun, 64,165
  right to take, 28,37
  treatment choice, 50
  unknown magnitude of, 40- 41,49
Risks, lifestyle, 156,165
Root-Bernstein, Robert,
    on alternative medicine, 165
Rousseau, Jean-Jacques, 149
Rule utilitarian, 20,62

Safe sex, 91
Salzman, Brie, on LSTs, 53
Schopenhauer, Arthur
  critique of Kant, 16,
    17,20-21,23,24
  on self-harm, 24,31
Schorr, Andrew, 27-28,29
  on drug access, 97,98
Science.medicine.aids, Internet,
    52,69
Seat belts, autonomy rights
    and, 11,12
Self, as end, 20-21
Self-determination
  homosexuals and, 9
  issue of, 7
  mistakes and, 52
  religious intrusion and, 14
Self-harm
  right to intervene, 9
  views on, 24-26
Sex discrimination, 9
Singer, Peter
  on altruistic behavior, 99
  on equality, 153
Slavery, arguments on,
    34-35,38-39,57
Smart, J. J. C., on Kant, 20

Smoking, autonomy rights and, 11
Social Contract, The, 149
Sonnabend, Joseph
  on Fischl/Richman trial,
    107-108,111,112-113
  on placebo use, 2-3
  on substandard treatment, 128
Southeast Asia, health care in, 150
Sovereign authority, concept
    of, 42-43
Spitzig, Patricia, 111
State, experimental drug regulation,
    42,51,133-138,140
Stinkhorn, 69
Strong paternalism, 23
  Beauchamp and Childress on, 72
  Feinberg on, 78-79
  Robert Young on, 74-81
"Substantially autonomous," 30,65
Suffering
  Kant on, 22
  relation to autonomy, 27
Suicide
  Australian right to, 9-10
  Kant on, 21-22
Sullivan, Lucy, AIDS transmission
    prevention, 32
Sunday Times (London),
  on Concorde trial, 133
Surrogate markers, 107,116-119,142
"Surrogate Markers in AIDS
  and Cancer Trials," 127
Survival interests
  AIDS patient's, 8
  devaluation of, 7
  Kantian view of, 17
  sacrifice of, 91,115

Tännsjö, Torbjörn
  on judgement, 53
  on physician's role, 101
Ten, C. L.
  decision-making knowledge, 66
  on intervention, 33

Ten, C. L. *(continued)*
  on risk taking, 50
  on slavery, 34
Terminal illness, experimental drugs
    and, 1,5-6
Terminus technicus, autonomy as, 10
Thalidomide scandal, 4
Theory and Practice of Autonomy,
    The, 56
"Therapeutic misconception,"
    97,128-129
Therapeutic privilege, paternalistic
    intervention, 58,60
Threakall, Susan, 121,124
Totalitarian health state, 32
Treatment
  research as, 106-107
  substandard, 104-105,
    127,140-141,145
Treatment Action Group (TAG), 53
Treatment Issues, Peptide T, 48
Trust
  AIDS patient, 104
  physician's role and, 102

"Unconventional medicine," 68
Understanding, autonomy element,
    65,66
United Nations
  autonomy, 10,149
  health care rights, 146
United States, health care in, 150
Universal Declaration of Human
    Rights, 146
U. S. Court of Appeals, on individual
    rights, 95
U. S. Infectious Disease Society, 42
Utilitarian counting game, 99
Utilitarianism
  J. S. Mill's philosophy and, 30-32
  Kantian philosophy and, 19,20
Utility, autonomy and, 30

Victorian AIDS council, AIDS
 treatment, 106
Viral load, as surrogate marker,
 117-118
Vitamins, regulation of, 70
Voldberding, Paul, on placebo use,
 120
Volenti non fit iniuria, 24,31
Voluntary participation
 clinical trials, 7,9,103,126,137
 mistakes and, 46
Vouchers, 153

Waiver, paternalistic intervention,
 58,59
Waldron, Jeremy, 147
Wall Street Journal, The, on FDA
 deregulation, 7
"War against drugs," 13
Warner, Wolfgang, on enrollment,
 136
Washington Post, The, on Retrovir,
 113

Wasting syndrome clinical trial,
 96,105-106
Weak paternalism, 23,33,38
 Beauchamp, Childress and Faden
  on, 64-74
 Gerald Dworkin on, 56-64
 Joel Fineberg on, 41-56
Welfare system, 147
Wellcome Trust, 133
Weller, Ian, Concorde trial, 132,133
Wicclair, Mark R., on competence
 standards, 51,55-56
World Health Organization (WHO)
 health care rights, 146
 patient's rights, 105

Young, Robert
 on autonomy, 10,14,74-81
 strong paternalism, 74-81

Zidovudine, 113,114,119,
 122,124,133
Zippelius, Reinhold, 146

Order Your Own Copy of
This Important Book for Your Personal Library!

ACCESS TO EXPERIMENTAL DRUGS IN TERMINAL ILLNESS
Ethical Issues

_____ in hardbound at $49.95 (ISBN: 0-7890-0563-8)

COST OF BOOKS_____

OUTSIDE USA/CANADA/
MEXICO: ADD 20%_____

POSTAGE & HANDLING_____
(US: $3.00 for first book & $1.25
for each additional book)
Outside US: $4.75 for first book
& $1.75 for each additional book)

SUBTOTAL_____

IN CANADA: ADD 7% GST_____

STATE TAX_____
(NY, OH & MN residents, please
add appropriate local sales tax)

FINAL TOTAL_____
(If paying in Canadian funds,
convert using the current
exchange rate. UNESCO
coupons welcome.)

☐ **BILL ME LATER:** ($5 service charge will be added)
(Bill-me option is good on US/Canada/Mexico orders only;
not good to jobbers, wholesalers, or subscription agencies.)

☐ Check here if billing address is different from
shipping address and attach purchase order and
billing address information.

Signature _____

☐ **PAYMENT ENCLOSED:** $_____

☐ **PLEASE CHARGE TO MY CREDIT CARD.**

☐ Visa ☐ MasterCard ☐ AmEx ☐ Discover
☐ Diner's Club

Account # _____

Exp. Date _____

Signature _____

Prices in US dollars and subject to change without notice.

NAME _____

INSTITUTION _____

ADDRESS _____

CITY _____

STATE/ZIP _____

COUNTRY _____ COUNTY (NY residents only) _____

TEL _____ FAX _____

E-MAIL_____
May we use your e-mail address for confirmations and other types of information? ☐ Yes   ☐ No

Order From Your Local Bookstore or Directly From
The Haworth Press, Inc.
10 Alice Street, Binghamton, New York 13904-1580 • USA
TELEPHONE: 1-800-HAWORTH (1-800-429-6784) / Outside US/Canada: (607) 722-5857
FAX: 1-800-895-0582 / Outside US/Canada: (607) 772-6362
E-mail: getinfo@haworthpressinc.com
PLEASE PHOTOCOPY THIS FORM FOR YOUR PERSONAL USE.

BOF96